PERGAMON INTERNATIONAL LIBRARY
of Science, Technology, Engineering and Social Studies
*The 1000-volume original paperback library in aid of education,
industrial training and the enjoyment of leisure.*
Publisher: Robert Maxwell, M.C.

TECHNIQUES IN BIOPRODUCTIVITY

AND

PHOTOSYNTHESIS

TECHNIQUES IN BIOPRODUCTIVITY
AND
PHOTOSYNTHESIS

Edited by

J. COOMBS and D. O. HALL

Sponsored by the United Nations Environment Programme

PERGAMON PRESS

OXFORD · NEW YORK · TORONTO · SYDNEY · PARIS · FRANKFURT

U.K.	Pergamon Press Ltd., Headington Hill Hall, Oxford OX3 0BW, England
U.S.A.	Pergamon Press Inc., Maxwell House, Fairview Park, Elmsford, New York 10523, U.S.A.
CANADA	Pergamon Press Canada Ltd., Suite 104, 150 Consumers Road, Willowdale, Ontario M2J 1P9, Canada
AUSTRALIA	Pergamon Press (Aust.) Pty. Ltd., P.O. Box 544, Potts Point, N.S.W. 2011, Australia
FRANCE	Pergamon Press SARL, 24 rue des Ecoles, 75240 Paris, Cedex 05, France
FEDERAL REPUBLIC OF GERMANY	Pergamon Press GmbH, 6242 Kronberg-Taunus, Hammerweg 6, Federal Republic of Germany

First edition 1982

British Library Cataloging in Publication Data

Techniques in bioproductivity and
    photosynthesis.—(Pergamon international
    library)
    1. Plant physiology
    2. Field crops
    3. Photosynthesis
    I. Coombs, J.   II. Hall, D. O.
    631.5   SB185.5   80-42235

ISBN 0 08 027382 3 hardcover
      0 08 027383 1 flexicover

The designations employed and the presentation of material in this publication do not imply the expression of any opinion on the part of the secretariat of the United Nations Environment Programme concerning the legal status of any country, territory, city or area or of its boundaries.

Printed in Great Britain by A. Wheaton & Co., Ltd., Exeter

This manual has been prepared from materials used in training courses and workshops under a project (FP/1303-78-02) of the United Nations Environment Programme and King's College, University of London with additional financial support and participation from the following organizations: United Nations Development Programme; United Nations Educational, Scientific and Cultural Organization; Commission of the European Communities; British Council; Commonwealth Foundation; Bundesministerium fur Wissenschaft und Forschung (Austria); Tate and Lyle Ltd (UK); National Science Foundation (U.S.A.); Madurai Kamaraj University (India); University of Nairobi (Kenya); Maize Research Institute (Yugoslavia).

Contributors to the manual:

J. Coombs,
Tate and Lyle Ltd.,
P.O. Box 68,
Reading,
RG6 2BX, U.K.

D. O. Hall,
King's College,
University of London,
68 Half Moon Lane,
London, SE24 9JF, U.K.

C. L. Beadle,
47 Davies Ave.,
Leeds,
LS8 1JZ, U.K.

H. R. Bolhàr,
Inst. Plant Physiol.,
University of Vienna,
Dr. Karl Lueger-Ring 1,
A-1010 Vienna, Austria

G. E. Edwards,
Dept. of Horticulture,
University of Wisconsin,
Madison,
Wisconsin 53706, U.S.A.

M. G. Guerrero,
Dept. of Biochemistry,
University of Sevilla,
Sevilla,
Spain

J.-E. Hällgren,
Dept. of Plant Physiol.,
University of Umeå,
S-90187 Umeå,
Sweden

G. Hind,
Dept. of Biology,
Brookhaven Nat. Lab.,
Upton, L.I., N.Y.,
11973, U.S.A.

P. J. Lea,
Dept. of Biochemistry,
Rothamsted Exp. Stat.,
Harpenden, Herts.,
AL5 2JQ, U.K.

R. Leegood,
Dept. of Botany,
University of Sheffield,
Sheffield,
S10 2TN, U.K.

S. P. Long,
Dept. of Biology,
University of Essex,
Colchester,
CO4 3SQ, U.K.

M. M. Ludlow,
C.S.I.R.O.,
Cunningham Laboratory,
Mill Road, St Lucia,
QLD 4067, Australia

H. Maske,
Institut Meereskunde,
University of Kiel,
Dusternbrooker Weg 20,
D-2300 Kiel, Germany

M. Reporter,
Charles F. Kettering Lab.,
150 E. S. College St.,
Yellow Springs,
Ohio, 45387, U.S.A.

L. L. Tieszen,
Dept. of Biology,
Augustana College,
Sioux Falls,
S.D. 57102, U.S.A.

A. Vonshak,
Dept. of Biology,
Sde Boker Campus,
P.O. Box 653,
Beersheva, 84120, Is.

D. A. Walker,
Dept. of Botany,
University of Sheffield,
Sheffield,
S10 2TN, U.K.

The following people have contributed to the training courses:

C. L. Beadle	(U.K.)	S. Krishnaswamy	(India)
H. R. Bolhàr	(Austria)	P. J. Lea	(U.K.)
P. Chartier	(France)	R. C. Leegood	(U.K.)
J. Coombs	(U.K.)	S. P. Long	(U.K.)
G. E. Edwards	(U.S.A.)	M. M. Ludlow	(Australia)
H. Egneus	(Sweden)	H. Maske	(Germany)
A. Gnanam	(India)	C. B. Osmond	(Australia)
M. G. Guerrero	(Spain)	C. Radenovic	(Yugoslavia)
D. O. Hall	(U.K.)	M. Reporter	(U.S.A.)
J.-E. Hällgren	(Sweden)	Z. Stankovic	(Yugoslavia)
B. Hesla	(U.S.A.)	L. L. Tieszen	(U.S.A.)
G. Hind	(U.S.A.)	A. Vonshak	(Israel)
U. Horstmann	(Germany)	Z. Vucinic	(Yugoslavia)
S. K. Imbamba	(Kenya)	D. A. Walker	(U.K.)
J. S. Kahn	(U.S.A.)		

PREFACE

This manual has evolved as a by-product of a scheme sponsored by the United Nations Environment Programme to provide training in the field and laboratory techniques associated with the measurement of plant productivity, with particular emphasis on photosynthesis. So far three training courses have been held, one in India, one in Kenya and one in Yugoslavia. Each course ran for 2 weeks and was attended by about twenty-five students from developing countries. The aim of the courses has been to train students to apply the best available and most relevant techniques to their own problems. Hence, they have been conducted in an environment similar to that which they will find in their own countries relying on local facilities, often using equipment manufactured or adapted from that available in the host institution. This publication is based on the content of the courses. It does not set out to be all-inclusive, or to be a lecturer's handbook, but rather reflects the interactions between the students and the lecturers on the courses, covering the areas of study in which the students have had the greatest interest or desire to learn.

The courses are obviously aimed at meeting a need, that of increasing knowledge of productivity of plant communities in warmer regions. The importance of this reflects the major problem facing many such countries—the rising cost of energy and its relevance to productivity in agriculture and forestry and to providing local sources of energy. This is particularly true of those areas which lack both fossil-fuel reserves and the industrial base to earn money to pay the ever-increasing oil bill. For such countries the only answer may be a greater reliance on indigenous energy sources such as geothermal, hydroelectric and solar. As far as many warmer countries are concerned the best, or in some cases the only, option appears to be the use of solar energy trapped by growing plants, i.e. biomass. The use of biomass as a fuel presents one major problem at present—the conflict between the production of food or the production of an energy crop in areas where land, water or other resources may be finite. Ironically, many of the countries which appear to offer the best opportunities for a biomass programme are those facing the greatest food shortages. This is, of course, the result of interaction between many factors, some agricultural, some social, some economic. However, in theory it should be possible to define the potential of any given area, crop or ecosystem to produce biomass, and to attach a numerical value to the amount of plant material which can be produced. In other words, it should be possible to **predict the potential productivity of agricultural, forestry or aquatic systems, in terms of the photosynthetic capacity of both existing crops and possible new plant species currently not being fully exploited. To do this requires that the plants are studied *in situ*,** rather than extrapolating from studies carried out in temperate conditions. For example, following the elucidation of the photosynthetic carbon-reduction cycle, many photosynthesis researchers considered that there was little left to discover about photosynthetic carbon metabolism. No one could have predicted the existence of C_4 plants which are adapted to sunnier and/or drier areas. In the same way the biology of nitrogen fixation in symbiotic associations in tropical legumes and grasses differs from that in many of the standard temperate crops used as experimental material for much of the work currently reported in the literature.

Without adequate training the scientist experimenting with new species, or in a new ecosystem, and obtaining results which differ from those in the textbook, may discard what could be a new discovery as important as that of C_4 photosynthesis. At a more practical level the worker using instruments in the field, to measure CO_2 fixation, light intensity or moisture stress or even carrying out analysis of crop growth by simple gravimetric means, may be faced with problems of calibration, technique or interpretation which lie well outside the scope

of his experience. The aim of these courses has been to enable a carefully chosen group of young people entering this area of work to interact with experienced researchers so that they can gain the skills to solve such problems as they encounter in their own work—and perhaps of even greater importance spread the knowledge that they have gained amongst their fellow countrymen.

J. Coombs
D. O. Hall

CONTENTS

INTRODUCTION

It is often possible to use plant material as a source of fuel and fibre and at the same time provide enough food in many of the warmer countries. Types of processes which might be used are summarized in Section 10, which also shows that the **products of photosynthesis** are about ten times the world's present total energy use. Furthermore, the standing biomass is comparable with proven reserves of fossil fuel. As far as land plants are concerned the major production occurs in the warmer regions. It is also in these regions that species with the highest rates of production are found.

In order to use the potential of the **biomass** to the full value considerable scientific and technological input will be required. The aims of any system designed for sustained use of plant material are:

(a) High yields.
(b) Low inputs.
(c) Use of all plant material.
(d) Use of process wastes.
(e) Maximum use of land, water, fertilizer, etc.
(f) Selection of plants for non-food as well as for food biomass.

In general, much **agriculture** in warmer countries is at the subsistence level. This is characterized by:

(a) Absence of cash inputs.
(b) Low crop yields.
(c) Decreasing soil nutrient content.
(d) Shifting cultivation, often with destruction of forest, followed by destruction of the soil.
(e) Production increases resulting from an expansion of the cultivated area with yields per hectare remaining static or falling.

In contrast agriculture in the developed countries has many inputs such as:

(a) Inorganic fertilizers.
(b) (Chemical) pest, disease and weed control.
(c) Mechanical cultivation and harvesting.
(d) Storage and process facilities.

If these inputs are available and put into subsistence-level farming, an exponential increase in production can be obtained. However, this is correlated with a decrease in **energy input/output ratios** (i.e. a higher energy requirement) and mechanization (i.e. a lower use of manpower). An alternative approach is to increase the scientific inputs into the system so that the natural resources of biological nitrogen fixation and recycling of organic material lead to increased soil fertility and higher yields. Other factors such as extremes of temperature, deficiencies of water, nutrients or light, disease and pests, also decrease yields, often through their effect on photosynthesis or on the plant's ability to carry out photosynthesis.

In theory about 6% of the solar energy falling on a given area could be converted to plant organic material. However, in practice, yields are consistent with conversion efficiencies well below this figure. In order to reach towards the theoretical value, and possibly to improve it, an understanding of photosynthesis and the techniques which have been developed in order to investigate it are of paramount importance.

So far the term **yield** has been used without

definition. Yields can be expressed in terms of amount of plant material produced in a given time (usually one year or one crop period) on the basis of land area. On the other hand, in consideration of plant material as an energy source, yields expressed in terms of amount produced per energy unit input, or per person, may be of equal importance.

The total photosynthate produced may be termed true biological yield. This will differ from the useful or economic yield which will be of smaller magnitude. The fraction used is known as the harvest index:

$$\text{Harvest index} = \frac{\text{economic yield}}{\text{biological yield}} \times 100$$

The growth rate and hence productivity of a crop is limited by the size of the assimilatory system, i.e. the leaf area of the crop at any time, and dependent on the net assimilation rate, i.e. the ability of this leaf area to fix CO_2 by photosynthesis. Therefore in simple terms:

$$CO_2 + H_2O \xrightarrow[\text{chlorophyll}]{\text{light}} (CH_2O)_n + O_2$$
$$\text{(Leaf area)}$$

Since light and CO_2 assimilation rates are important limiting factors, accurate measurement is essential. This is covered in Part I. In the above equation the dry matter produced is represented by $(CH_2O)_n$ or carbohydrate. Of course in the plant carbohydrate is metabolised further to proteins, lipids, lignin, etc. The direct determination of dry matter measured at the end of an experimental period (described in Section 1) gives a value which can be misleading as it represents only the difference between what has been produced and what has been lost. It also gives little information about the underlying growth process. To understand such processes, an understanding of both the physiology and biochemistry of photosynthesis is required. Some aspects of these subjects are dealt with here particularly on laboratory techniques used for isolation of chloroplasts and enzymes from plant tissue. In addition to methods applicable to photosynthetic CO_2 assimilation, methods for in-

vestigating nitrogen fixation are detailed—the importance of the nitrogen status of plants in relation to photosynthetic productivity is often ignored.

Experimentation

In order to obtain meaningful information accurate observations are required which in turn means experimentation, i.e.

(a) Design.
(b) Perform.
(c) Observe.
(d) Record.
(e) Conclude.

The importance of these elements are stressed throughout the course. In particular the following aspects should be noted:

(a) *Design*   (i) Define question to be answered.
                (ii) Choose adequate healthy material—identify it.
                (iii) If using material over a period of time be systematic, record all inputs, note any disease or pests or treatments.
                (iv) Have adequate controls.
                (v) Be simple—answer one point at a time.

(b) *Perform*   (i) Be accurate.
                (ii) Avoid contamination.
                (iii) Avoid stresses or artifacts.

(c) *Observe*   (i) Be critical.
                (ii) Be honest.
                (iii) Be accurate.
                (iv) Be objective.

(d) *Record*   (i) Be accurate.
                (ii) Be systematic.
                (iii) Be thorough.

(e) *Conclude* Be objective. Is your conclusion justified or have you been biased towards a preconceived idea?

Recording and presenting your results

It is essential that a clear and concise account is kept of all your experimental results and conclusions. It is suggested that you keep two notebooks, a small one for recording observations as they are made, for calculating, for example, how much of a given chemical to weigh out when making solutions, and a larger notebook (about 21×30 cm, A4 size) in which all experiments should be fully written up. The experimental reports are examined at the end of the course and taken into account when final evaluations are made.

Experiments should be set out under the following headings: *title* (complete with a date and experiment code number which should also be used on each page of the small notebook when primary data is being recorded); a description of the *aims* of the experiment; the plant *material* used; the *methods* used; the *results* (numerical) plus a note of any unusual delays or occurrences which might have contributed to spurious results; *calculations* of results derived from the primary data (e.g. leaf area, rates of reaction, concentration of chlorophyll); *conclusions*. In the conclusions it should be noted whether these have answered the questions as detailed in the aims of the experiment, suggestions for further or better experiments should also be given.

In recording results care should be taken in respect to the following:

(a) *Replication*, a single weighing may be sufficient but most other measurements should be taken at least twice. At least two replicate determinations, using material from a similar source, should be made. In general, agreement should be such that they do not vary by more than 5%. If the variation is larger, sampling must be repeated; if continued variation is observed, check the method being used.

(b) *Significance*, do not record your results to a greater accuracy than the variation between replicates justify, e.g. two weighings of 11.349 and 12.016 would be recorded as 11.7 rather than 3 places of decimals.

(c) *Statistics*, do not give a mean and standard error when you only have a few results; your calculator may give a result but it is not very meaningful unless you are dealing with samples of 30 or more. However, if you find your results appear to fall on a straight line do use a calculator to obtain the "best fit" (least squares), slope and intercept.

It is essential that all experimental measurements are completed and that all data is treated as fully as possible. For instance, results of measurements of rates of CO_2 assimilation should be combined with determinations of leaf area, fresh weight, dry weight and chlorophyll content to give rates of photosynthesis in terms of leaf area, weight or chlorophyll content for a given light intensity.

Again, in biochemical experiments results should be treated in terms of weight, chlorophyll or protein content. For enzyme experiments results should be expressed in terms of the amount of substrate converted in a given time by a given amount of protein.

Please enjoy the work and derive maximum benefit from it!

PART I

WHOLE PLANT PHOTOSYNTHESIS

WHOLE PLANT PHOTOSYNTHESIS AND PRODUCTIVITY

1.1. INTRODUCTION

by S. P. LONG

The first section of your manual is concerned with the measurement and analysis of productivity and photosynthesis of intact higher plant tissues, i.e. leaves, whole plants and stands.

When dealing with the efficiency of light-energy conversion into biomass in higher plants concern commonly centres on such questions as why is a genotype more productive in one environment than in another or what are the limitations to productivity for a given genotype in a given environment. A common mistake in scientific approach to such a problem is first to look at the isolated parts rather than the whole. For example, in analysing the question of why increased salinity decreases the productivity of a crop variety it would be pointless to first look at changes in single leaf rates of CO_2 assimilation, isolated RuBP carboxylase activity nor amounts of assimilated $^{14}CO_2$ that are incorporated into different compounds since even if salinity-induced changes are found neither process is necessarily limiting productivity. We cannot even be sure that a reduction in productivity has anything to do with an effect of salinity on the photosynthetic apparatus, it could equally well be an effect on leaf area or canopy structure causing changes in amounts of intercepted solar radiation. Thus, analysis of the whole plant forms the first logical step in analysing the basis of limitations to productivity in higher plants. This is not to say that subcellular studies have no role in analysing photosynthetic efficiency as clearly they have a role in improving our knowledge of the photosyn-thetic process and its control, but some thought should be given to whether or not subcellular studies are immediately relevant to the problem that you are trying to solve. Figure I.1 illustrates a logical sequence of steps for reducing the question of what factors are limiting the efficiency of light-energy conversion into biomass in a given crop, stand or single plant. However, even before such questions can be tackled it is essential that we can define the environment and productivity of our plants.

Section 1.2 deals with measurement of the aerial environment of the plant, since it is this rather than the soil or edaphic environment which directly influences photosynthetic rates of individual leaves in the short term. The soil can of course be equally important in the longer term. Chemical analysis of soils and plant tissues is not dealt with in this section; three useful sources are Chapman (1976), Epstein (1972) and Hesse (1971). Measurement of the productivity of stands is considered in Section 1.3. Data on biomass changes with time can be used for Growth Analysis, a procedure which allows partitioning of limitations to production between leaf area and assimilatory capacity (Section 1.4). Hunt (1978) provides an invaluable introduction to this topic, Květ (1971) and Evans (1972) provide more detailed reference sources.

Traditionally plant productivity has been measured by dry-weight changes measured by destructive harvesting at intervals of days, weeks or even months. However, measurement of CO_2 exchange allows instantaneous estimation of productivity on a minute by minute basis. Thus effects of short-term changes in the environment on productivity can be detected.

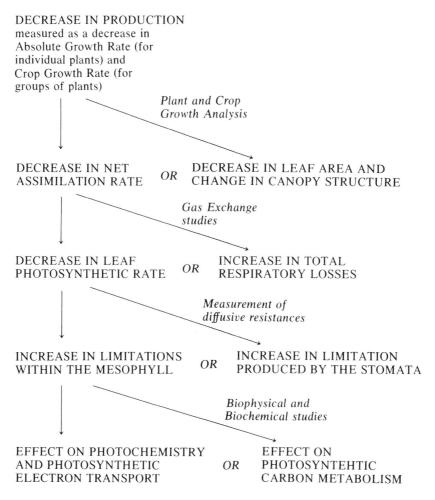

FIG I.1. A reductive analysis of factors limiting plant/crop productivity.

Gas-exchange systems are most commonly used in laboratory-controlled environment studies (Section 1.5), although they have also been used in the field and simple portable systems using $^{14}CO_2$ are described here (Section 1.6). However, measurement of exchange of CO_2 and water vapour provides more than instantaneous estimates of photosynthesis and transpiration. The relative importance of different limitations to photosynthetic efficiency within the canopy and the leaf can be assessed by the use of a resistance analogue model (Section 1.5) which can be evaluated from gas-exchange measurements.

Measurement of stomata as a limitation to CO_2 assimilation and transpiration is considered

in Section 1.7. Finally, as far as whole plant photosynthesis is concerned, Section 2 deals with the relationship of the organization of leaves in the shoot and cells within the leaf to the efficiency of light-energy conversion in photosynthesis.

Bibliography and Further Reading

ALLEN, S. E. (ed.) (1974) *Chemical Analysis of Ecological Materials* (Blackwell: Oxford).

CHAPMAN, S. B. (ed.) (1976) *Methods in Plant Ecology.* (Blackwell: Oxford).

EPSTEIN, E. (1972) *Mineral Nutrition of Plants: Principles and Perspectives.* (Wiley: N.Y.)

EVANS, G. C. (1972) *The Quantitative Analysis of Plant Growth.* (Blackwell: Oxford.)

HESSE, P. R. (1971) *A Textbook of Soil Chemical Analysis.* (John Murray.)

HUNT, R. (1978) *Plant Growth Analysis (Studies in Biology 98).* (Edward Arnold: London.)

KVĚT, J. (1971) see Šesták *et al.* (eds.) (1971).

MONTEITH, J. L. (1973) *Principles of Environmental Physics.* (Edward Arnold: London.)

MONTEITH, J. L. (ed.) (1975) *Vegetation and the Atmosphere,* 2 vols. (Academic Press: London.)

ŠESTÁK, Z., ČATSKÝ, J. and JARVIS, P. G. (eds.) (1971) *Plant Photosynthetic Production. Manual of Methods.* (Dr. W. Junk: The Hague.)

1.2. MEASUREMENT OF SOLAR RADIATION, TEMPERATURE AND HUMIDITY

by M. M. LUDLOW

1.2.1. Measurement of Solar Radiation

1.2.1.1. Introduction

Solar radiation is the driving force of photosynthesis and plant growth. The amount of radiation received sets the upper limit of biomass production and it determines the ecological distribution of plants by influencing their energy balance and temperature. Whether the energy of solar radiation is being converted to heat, to electricity using solar cells, or to biomass, it is necessary to measure the total amount of energy or the energy within specific wavelength bands to study and understand the conversion processes. The following presentation defines solar radiation, and describes the theory and practice of its measurement.

1.2.1.2. Solar Radiation

Solar radiation reaching the earth's surface can be separated into two components based on whether the radiation comes directly from the sun (direct) or whether it is scattered or reflected by clouds and the atmosphere (diffuse). Irrespective of whether it is direct or diffuse the radiation can be separated into two wavelength bands.

Short-wave radiation   <3000 nm
Thermal radiation   > 3000 nm

Shortwave radiation can be further broken down into a number of narrower bands which vary in their effects on plant growth as shown in Tables 1.1 and 1.2. Radiation in the visible range (400–700 nm) is of particular interest because those wavelengths are photosynthetically active (PAR). The spectral bands of solar radiation are usually characterized in terms of energy. This is adequate for all except the visible range.

Table 1.1. *Spectral bands of solar radiation: their effects on plants and instruments to measure them*

Name	Wavelength range (nm)	Effects on plant growth	Instrument
Ultraviolet	<280	Rapidly killed	Thermopile with special filters
	280–315	Detrimental	
	315–400	Morphogenetic	
Visible	400–700	Photosynthetic	Quantum flux meter
Red	610–700	Morphogenetic (phytochrome—red)	Thermopile with special filters
Far-red	700–800	Morphogenetic (phytochrome —far-red)	
Short-wave	300–3000	Photosynthetic, morphogenetic, energy balance	Glass-covered thermopile
Terrestrial (thermal)	$3 \times 10^3 - 10^5$	Energy balance	Filtered thermopile
Total radiation Net radiation	$300 - 10^5$	Energy balance, water relationships	Polythene-covered thermopile

Table 1.2. *The main spectral regions of physiological importance to plants (after Kubin, 1971)*

Spectral region	Character of absorption	Physiological effect
Infra-red		
>1000 nm	By water in tissues	Without any specific effect on photochemical and biochemical processes; converted into heat
1000–720 nm	Slight	Stimulating elongation
Photosynthetically active radiation		
720–610 nm	Very strong, by chlorophylls	Large effect on photosynthesis and photoperiodism
610–510 nm	Somewhat less	Small effect on photosynthesis; small morphogenetic effect
510–400 nm	Very strong, by chlorophylls and carotenoids	Large effect on photosynthesis; large morphogenetic effect
Ultraviolet		
400–315	By chlorophylls and protoplasm	Without any specific effect, small effect on photosynthesis
315–280 nm	By protoplasm	Large morphogenetic effect; stimulating some biosynthesis; large effect on physiological processes
<280 nm	By protoplasm	Lethal in large doses

Because photosynthesis is a photochemical process, the photon flux is more meaningful than energy flux. The photon flux is determined by the number of photons, and is independent of the energy per photon.

1.2.1.3. How to Measure Solar Radiation

There are a number of requirements for instruments to measure the various spectral bands of solar radiation. These are in addition to the normal requirements of reliability, reproducibility, accuracy, stability and insensitivity to other environmental factors.

1.2.1.3(i). RESPONSIVENESS TO THE WAVELENGTHS BEING STUDIED

This is achieved by either sensors which are insensitive to other wavelengths, or by sensors which are sensitive to all wavelengths but where provision is made for unwanted wavelengths to be filtered out.

1.2.1.3(ii). RESPONSE OF THE SENSOR MUST BE SIMILAR TO THAT OF THE PLANT

For example, if photosynthetic quantum flux is being measured, the sensor should be insensitive to wavelengths shorter than 400 nm and longer than 700 nm and should respond to light quanta in a manner similar to photosynthesis (discussed later under Quantum Flux).

1.2.1.3(iii). COSINE RESPONSE

By convention radiant flux is usually measured with the sensor held horizontal and the maximum value occurs when the sun is overhead. However, when the radiation is at angles less than 90°, the amount of energy per unit area will be less according to the cosine of the angle of incidence (α):

$$I = I_0 \cos \alpha$$

where I_0 and I are, respectively, the irradiance on a surface normal to the sun's rays and on a surface with an angle α between the normal to

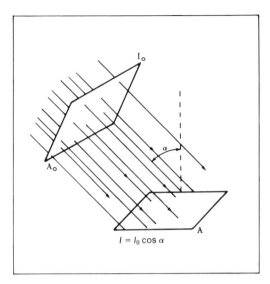

FIG. 1.1. The relationship between the irradiance (I_0) on a surface normal to the sun's rays and on a surface with an angle α between a normal to the surface and the rays. The area of A is greater than A_0 and hence $I < I_0$.

the surface and the rays (Fig. 1.1.). Sensors should be designed so that they obey this cosine law as closely as possible.

Different methods are used to achieve a good cosine response, but a common way is with a raised white perspex diffuser on top of the sensor.

1.2.1.4. Instruments

Total and net radiation are measured with an instrument called a radiometer which incorporates a thermopile. A thermopile is a black surface composed of many thermocouple junctions. The black surface absorbs all wavelengths of radiation and heats up. The temperature reached, relative to some reference temperature inside the instrument is proportional to the absorbed energy. The thermopile is covered with a polyethylene dome which is transparent to all wavelengths and which protects the sensor from dust and water. Total radiation is that falling on a horizontal surface, whereas net is total minus that reflected or radiated from the surface. Thus a net radiometer has two thermopiles back to back, one responding to the incident radiation and the other to that coming away from the

surface under study, and their outputs are subtracted electrically. Net radiation is commonly measured in energy and water-balance studies.

Thermal radiation is measured using a similar type of instrument but with a filter which excludes short-wave radiation.

Short-wave radiation conversely is measured using a thermopile similar to that used for total radiation except it has a glass dome to exclude thermal radiation (>3000 nm). A difficulty with this arrangement, however, is that the glass dome having absorbed the thermal radiation, heats up, and therefore radiates again inwards towards the sensor. This is overcome by using a double glass dome; for example, as in Kipp or Eppley solarimeters. Short-wave radiation measured with solarimeters is called irradiance and it has units of Wm^{-2}.

Solarimeters are used for routine meteorological measurements of solar radiation but they require an electronic integrator and mains power to give daily totals. They are also fairly expensive and this has prevented many people from making such measurements. However, CSIRO (Rauchfuss Instruments, Melbourne, Vic., Australia) has developed a cheaper solarimeter based on silicon cells rather than a thermopile. It does not require mains power and it gives integrated values of solar radiation.

Photosynthetically Active Radiation

The amount of energy in the visible range 400–700 nm (photosynthetically active radiation—PAR) is usually measured using a solarimeter with associated filters. It is difficult to get a filter which cuts off sharply at 700 nm whereas filters which remove the visible are available. Therefore two instruments have been used, one measuring total short-wave and the other with a filter to eliminate the visible. The visible is then found by difference. PAR is irradiance and the unit is Wm^{-2}.

Quantum Flux

Quantum flux is the most meaningful measurement that can be made for photosynthesis and growth studies. Because photosyn-

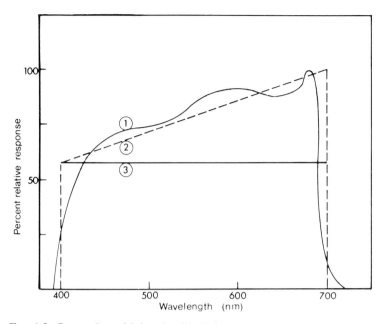

FIG. 1.2. Spectral sensitivity of an ideal photon response sensor in terms of photon numbers (solid horizontal line 3) and in terms of energy (broken line 2). The spectral response of a Lambda quantum sensor (solid curve 1) is also shown. Adapted from McPherson (1969) and Biggs *et al.* (1971).

thesis is influenced by the number of photons rather than their energy, the ideal quantum response for a sensor to measure quantum flux is one which is equally sensitive to photons between 400 and 700 nm (i.e. the flat response, 3, in Fig. 1.2). However, photoelectric cells which are used in quantum flux sensors measure energy rather quanta. Moreover, the energy per quanta decreases with wavelength. Therefore the ideal quantum response of sensors based on such cells is shown by the broken line (2) in Fig. 1.2. In other words, the broken line (2) is the spectral response of the sensor adjusted for the different energy levels at the blue and red ends of the spectrum. The Lambda quantum flux sensor which has been available for about 10 years is based on a silicon cell and a number of filters and has a response similar to the ideal (1 in Fig. 1.2). It is readily available and reasonably priced. The Lambda sensor also has a good cosine response. Thus there is now no excuse why adequate measurements of quantum flux cannot be made.

Light Intensity

Several types of instruments have been used to measure "light intensity". There have been a number of inaccuracies with both terminology and instrumentation. What is measured with a foot-candle or lux meter (such as a selenium cell) is *not* light intensity, but *illuminance* which is defined as luminous flux density intercepted per unit area and which has photometric units of foot-candles or lux.

A number of sensors have been used to measure illuminance (Fig. 1.3). The selenium cell and the selenium cell tailored for the response of the human eye were the most common types of cell employed even though they have low sensitivity in the blue and red ends of the spectrum. However, they do have a fairly sharp cut-off at 700 nm. The silicon cell approaches the ideal quantum response but it has no cut-off at 700 nm. In contrast the cadmium sulphide cell is sensitive to wavelengths greater than 700 nm.

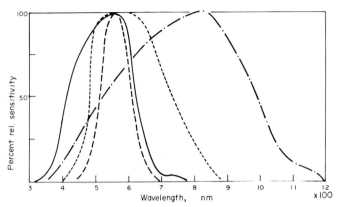

FIG. 1.3. Spectral sensitivities of various sensors: selenium cell (——); selenium cell corrected to match the sensitivity of the human eye (– – – –); silicon solar cell (–·–); and cadmium sulphide (------). Adapted from Szeicz (1968).

1.2.1.5. Special Problems Inside Canopies

Measurements of solar radiation above plant canopies are relatively simple as long as the matters discussed above are appreciated. Measurements within canopies, however, introduce a number of problems:

1.2.1.5(i). CHANGES IN SPECTRAL COMPOSITION

The spectral composition of solar radiation changes when it passes through chlorophyll-bearing tissue and, to a lesser extent, when it is reflected. Within the visible range, the proportion of red and blue decreases whereas the proportion of green increases but, more importantly, the proportion of near infra-red (700–800 nm) increases relative to the visible (Fig. 1.4). In the past we have been preoccupied with changes in the visible when really the greatest problem arises from the increase in near infra-red. When a cadmium sulphide cell which has a marked sensitivity to these wavelengths is used the amount of PAR inside canopies is greatly overestimated. Of the more readily obtained sensors, the selenium cell would appear to be satisfactory from this point of view. However, there is no substitute for the quantum flux meter which is untroubled by the change in quality here, or with artificial light sources.

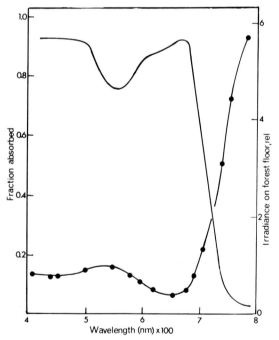

FIG. 1.4. Fractional absorption (absorptance) spectrum of an *Alocasia* leaf (upper curve), and the spectral distribution of diffuse radiant energy (lower curve) reaching the floor of a rain forest in south-eastern Queensland, Australia (Björkman and Ludlow, 1972).

1.2.1.5(ii). SPATIAL AVERAGING

There is a great variability of solar energy within plant communities, varying from sunflecks to· deep shade. Therefore a few spot measurements are likely to give bad estimates of the "average" value. Apart from the obvious approach of taking many spot readings which is difficult if incident radiation is changing, there have been two approaches to solving this problem.

1. *Linear probe* which consists of a number of sensors on a probe, or a long continuous sensor, both giving an "average value".
2. *A moving sensor.* A sensor is mounted on a wire or track and moves through the vegetation and integrates radiation along its path.

1.2.1.5(iii). PHYSICAL AND LOGISTIC PROBLEMS

The moving sensor works well in large crops such as maize but is more difficult in low-growing crops and pastures. Also it is large and cumbersome, and cannot be used to make a large number of measurements in different positions. The linear probe is a good compromise. It gives some spatial averaging, a large number of measurements can be made quickly, and dimensions are such that little disturbance is caused to the vegetation.

Changes in incident radiation while measurements are being made at various levels within the canopy is another logistic problem. Ideally measurements should be done on perfectly clear or uniformly overcast days. However, if this is not practicable the problem is partially overcome by expressing the values at each level relative to the incident radiation. A series of relative values at different levels in the canopy plotted against height is called a light profile or radiation profile. This is usually compared with leaf-area profiles when the light relations of crops are being studied.

1.2.1.5(iv). CONCLUSIONS

Linear probes composed of a number of sensors (Muchow and Kerven, 1977; Davis, Ludlow and Gibbs, 1980) or a continuous sensor (Lambda linear sensor—Li Cor Pty Ltd., Lin-coln, Nebraska, U.S.A.) which measure quantum flux overcome most of the problems encountered within plant canopies, especially when results are expressed relative to incident radiation.

Bibliography and Further Reading

ANDERSON, M. C. (1971) Radiation and crop structure. In: *Plant Photosynthetic Production, Manual of Methods* (Eds. Z. ŠESTÁK, J. ČATSKÝ and P. G. JARVIS), pp. 412–466. (Dr. W. Junk: The Hague.)

BAINNRIDGE, R., EVANS, G. C. and RACKHAM, O. (Eds.) (1966) *Light as an Ecological Factor.* (Blackwell Sci. Pub.: Oxford.)

BIGGS, W. W., EDISON, A. R., EASTIN, J. D., BROWN, K. W., MARANVILLE, J. W. and CLEGG, M. D. (1971) Photosynthesis light sensor and meter. *Ecology* 52, 125–131.

BJÖRKMAN, O. and LUDLOW, M. M. (1972) Characterisation of the light climate on the floor of a Queensland rainforest. *Yb. Carnegie Institute Wash.* 71, 85–94.

CALDWELL, M. M. (1971) Solar UV radiation and the growth and development of higher plants. In: *Photophysiology*, Vol. 6, Chap. 4, pp. 131–177. (Academic Press: New York.)

DAVIS, R., LUDLOW, M. M. and GIBBS, G. (1980) An instrument for measuring light profiles in pasture canopies. CSIRO Division of Tropical Crops and Pastures. Tech. Memo. No. 19.

DRUMMOND, A. J. (1965) Techniques for the measurement of solar and terrestrial radiation fluxes in plant biological research: a review with special reference to the arid zones. In: *Methodology of Plant Eco-physiology* (Ed. F. E. ECKARDT), pp. 13–27. (UNESCO: Paris.)

GATES, D. M. (1962) *Energy Exchange in the Biosphere.* (Harper & Row: New York.)

KUBIN, S. (1971) Measurement of radiant energy. In: *Plant Photosynthetic Production, Manual of Methods* (Eds. Z. ŠESTÁK, J. ČATSKÝ and P. G. JARVIS), pp. 702–765. (Dr. W. Junk: The Hague.)

MCCREE, K. J. (1972) Test of current definitions of photosynthetic radiation against leaf photosynthesis data. *Agric. Met.* 10, 443–453.

MCPHERSON, H. G. (1969) Photocell-filter combinations for measuring photosynthetically active radiation. *Agric. Met.* 6, 347–356.

MUCHOW, R. C. and KERVEN, G. L. (1977) A low cost instrument for measurement of photosynthetically active radiation in field canopies. *Agric. Met.* 18, 187–195.

PLATT, R. B. and GRIFFITH, J. F. (1969) *Environmental Measurement and Interpretation* (Reinhold: New York.)

RICHARDSON, J. A. (1964) *Physics in Botany.* (Pitman & Sons: London.)

ROSE, C. W. (1966) *Agricultural Physics* (Pergamon Press: Oxford.)

STRINGER, E. T. (1972) *Techniques of Climatology.* (W. H. Freeman & Comp.: San Francisco.)

SZEICZ, G. (1968) Measurement of radiant energy. In: *The Measurement of Environmental Factors in Terrestrial Ecology* (Ed. R. M. WADSWORTH), pp. 109–130. (Blackwell Sci. Pub.: Oxford.)

1.2.2. Temperature Measurement

1.2.2.1. Introduction

Temperature influences rates of biochemical reactions; for example, rates double or triple with each 10°C rise in temperature. Temperature also affects rates of plant growth and development, such as rates of leaf production and expansion, and flowering (Sutcliffe, 1978). Low temperature and frost at one extreme, and high temperature at the other, affect the survival and ecological distribution of plants. In plant biology we are primarily interested in the temperature of air, plants, plant parts and soil. Units for temperature are degrees Celsius (°C) or degrees Kelvin (K = 273 + °C). Temperature in K is also called the absolute temperature.

1.2.2.2. Principles of Temperature Measurement

Temperature is not measured directly, as is mass, but indirectly by its influence on other physical properties; for example, the change in volume of mercury in a mercury-in-glass thermometer or the change in electrical resistance of a platinum resistance thermometer. Various principles of temperature measurement are outlined below.

1.2.2.2(i). CHANGE IN PHYSICAL DIMENSIONS

This is the most common method employed and is the basis of the simplest and most reliable techniques. In liquid-in-glass thermometers a change in temperature is registered as the change in volume of the fluid (e.g. mercury-in-glass thermometers). Another example is the bimetallic strip used in thermographs. It is composed of two metals with different coefficients of expansion

bonded together and formed into an incomplete circle. A change in temperature causes the strip to deform and this is registered on a dial or chart through a system of levers.

1.2.2.2(ii). CHANGE IN ELECTRICAL PROPERTIES

A change in temperature results in a change in electrical properties of various sensors. The resistance of both platinum resistance thermometers and thermistors change with temperature. On the other hand, the voltage drop across a diode varies with temperature. The flow of electrons between two dissimilar metals in a thermocouple also changes with temperature. Thermocouples are widely used in biology because of their small size, simplicity and cheapness (Richardson, 1964; Platt and Griffith, 1969). When two thermocouple junctions are joined as shown below the voltage (v) is proportional to the difference in temperature ($t_1 - t_2$).

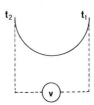

$$v \propto (t_1 - t_2)$$

or

$$v = k(t_1 - t_2)$$

where k is the temperature coefficient depending on the nature of the two metals; for example, the value for copper/constantan thermocouples is about $40\ \mu\text{V}\ °\text{C}^{-1}$.

If the temperature of one junction is known (the reference temperature, t_2), the temperature of the other can be calculated.

$$t_1 = \frac{v}{k} + t_2.$$

Melting ice (0°C) or water at a known temperature (measured with a mercury-in-glass thermometer) in a stirred container is commonly used to provide the reference temperature. Soil temperature at a depth of 1 m is quite stable and can be used as a reference if its temperature is measured with a thermometer.

Sometimes an electronic reference is used instead of a reference thermocouple. The Wescor (Logan, Utah, U.S.A.) thermocouple thermometer (models TH-50 and TH-60) is one such instrument which only requires one thermocouple to measure temperatures.

A range of metals can be used to make thermocouples but copper and constantan are widely used because they are cheap, convenient and do not rust or corrode (Perrier, 1971). Thermocouple wire comes in a wide range of diameters from as fine as human hair to as thick as a pencil.

1.2.2.2(iii). CHANGE IN RADIATIVE PROPERTIES

All surfaces emit energy (Q) in proportion to their absolute temperature (T) according to the Stefan–Boltzmann Law:

$$Q = \epsilon \sigma T^4$$

where σ is the Stefan–Boltzmann constant $(5.67 \times 10^{-8}\ W\ m^{-2}\ K^{-4})$ and ϵ is the emissivity, the value of which varies from 1 for a perfect blackbody radiator to <1 for most other surfaces. The emissivity for most leaves approaches 1 because of their high water content. However, emissivity varies among soil types depending upon texture, colour and moisture content.

Infra-red thermometry, thermography and thermoluminescence all use the energy emitted from surfaces as a means of measuring surface temperature (Perrier, 1971). These are called non-contact methods because they do not require contact with the surface being measured. They have a considerable advantage over contact methods which can alter surface temperature during measurement.

Infra-red thermometers measure the energy being emitted in a particular wavelength band $(8-13\ \mu m)$ which is directly related to the surface temperature. On the other hand, infra-red thermography measures thermal radiation emitted by a body when the surface is scanned several times per second. A third method for measuring surface temperature is thermoluminescence. Thin layers of a luminescent compound are deposited on a surface and emit visible radiation, some properties of which vary with temperature. All three non-contact methods are expensive, complex and of limited accuracy for absolute temperature measurements. However, the technology of infra-red thermometry is developing rapidly, and in the future we may have relatively cheap, accurate, non-contact thermometers.

1.2.2.3. General Problems of Temperature Measurement

1.2.2.3(i). NEED TO MEASURE THE MOST APPROPRIATE TEMPERATURE

Meteorological records give good long-term average temperatures for a particular area. However, temperatures at places of different topography, altitude and aspect may differ greatly from those recorded at the meteorological site. Therefore temperature should be recorded at each agronomic experimental site or ecological situation if they are not close to meteorological stations. In addition, the most meaningful plant temperature should be measured. Terrestrial minimum temperature is far more meaningful in frost studies than air temperature measured in a screen 2 m above the ground. Moreover, temperatures of leaves, fruits and apices are more useful than air temperature in understanding rates of biochemical reactions, plant growth and development.

1.2.2.3(ii). ACHIEVING GOOD CONTACT WITH THE OBJECT BEING MEASURED

A prequisite of all contact methods is to make good contact between the sensor and the object being measured, while ensuring that the presence of the sensor does not influence the temperature. Thermocouples, because of their small size, are often used to measure surface temperatures where good contact is particularly important. They are either held to the surface by surgical adhesive tape or a spring. Care should be taken to achieve maximum contact with the surface and to reduce the influence of the surrounding air. Non-contact methods do not suffer from this problem and are, therefore, particularly useful for measuring surface temperature. They can also be used to give an average temperature of a complex surface, such as a plant canopy.

1.2.2.3(iii). RADIATION SHIELDING

When measurements are being made in the presence of large fluxes of radiant energy such as in sunlight or near the soil surface, the temperature sensor must be shielded from this energy. Shielding is usually achieved by one or two highly reflective covers, which may also be ventilated to remove the absorbed energy.

1.2.2.4. Measurement of Air, Plant and Soil Temperature

1.2.2.4(i). AIR TEMPERATURE

Air temperature is measured in meteorological screens, which are shielded from radiation and ventilated, with mercury-in-glass or alcohol-in-glass thermometers, or with thermographs. These instruments are relatively cheap, reliable and accurate. Air temperature can also be measured with platinum resistance thermometers, thermistors, diodes and thermocouples as long as they are shielded from radiation and ventilated.

1.2.2.4(ii). PLANT TEMPERATURE

The temperature of plant organs such as fruits, stems or roots which have some mass and heat capacity can be measured with glass thermometers, platinum resistance thermometers, thermistors, diodes and thermocouples. Contact methods which use small sensors such as thermocouples or thermistors are suitable for measuring surface temperatures of leaves, fruits and stems. However, non-contact methods such as infra-red thermometry are preferable for surface temperature measurement.

The temperatures of plant organs are more useful than air temperature in understanding the influence of temperature on rates of biochemical and physiological processes and the influence of high and low temperature on plant survival. Leaves and other plant organs can be cooler or warmer than the surrounding air depending on the radiation environment (sunlight or cold clear skies at night), the windspeed, and their colour, size, shape and water status. Evaporating leaves, and leaves radiating energy to cool, clear skies, are often cooler than air temperature, whereas water-stressed leaves and alpine plants in sunlight are often warmer than air temperature. Leaf and air temperature will be closest in small leaves and in high winds.

1.2.2.4(iii). SOIL TEMPERATURE

Measuring the temperature of the soil surface has the same problems as measuring plant surface temperature. Non-contact methods are preferable as long as emissivity is known. Alternatively, shielded thermocouples or thermistors can be used. Temperature of deeper soil layers can be measured with glass thermometers (as in routine meteorological stations), thermistors, thermocouples, diodes and platinum resistance thermometers when good contact is made with the soil. Inorganic ions in the soil can cause some problems with some electric sensors. These problems are overcome by coating the sensors with a thin layer of epoxy resin.

Soil temperature decreases with depth from the surface during the day, and the reverse situation exists at night. Thus there is a marked diurnal variation in the temperature of the soil surface, but the magnitude of this variation declines with depth.

Bibliography and Further Reading

CAMPBELL, C. S. (1977) *An Introduction to Environmental Biophysics.* (Springer-Verlag: New York.)

MONTEITH, J. L. (1973) *Principles of Environmental Physics.* (Edward Arnold: London.)

PERRIER, A. (1971) Leaf temperature measurement. In: *Plant Photosynthetic Production, a Manual of Methods* (Ed. Z. ŠESTÁK, J. ČATSKÝ and P. G. JARVIS), pp. 632–671. (Dr. W. Junk: The Hague.)

PLATT, R. B., and GRIFFITH, J. F. (1969) *Environmental Measurement and Interpretation.* (Reinhold: New York.)

RICHARDSON, J. A. (1964) *Physics in Botany.* (Pitman & Sons: London.)

SUTCLIFFE, J. (1978) *Plants and Temperature. Studies in Biology* No. 86. (Edward Arnold: London.)

1.2.3. Measurement of Humidity

1.2.3.1. Introduction

The water vapour content, or humidity, is a measure of the dryness of air. It is an important determinant of rates of evaporation (E) and transpiration (T) which can be described by an analogue of Ohm's Law:

$$E(\text{or } T) = \frac{e_l - e_a}{r} \qquad (1.2)$$

where e_l and e_a are, respectively, the vapour pressure of a leaf or other evaporating surface and the vapour pressure of the air, and r is the resistance to water-vapour transfer from the evaporating surface to the air.

Recently it has been shown that stomata of some species respond directly to the humidity of the air, independently of the leaf water status. Thus stomata close in dry air, restricting exchange of carbon dioxide and water vapour, and thus possibly reducing growth.

For these two reasons, at least, humidity should be recorded routinely in meteorological stations and at specific agronomic and ecological sites during experiments. Before discussing methods for measuring humidity it is necessary to understand the physics of water vapour.

1.2.3.2. Physics of Water Vapour

Water vapour is a gas which exerts a partial pressure in air. The pressure exerted by water vapour in saturated air (*Saturated Vapour Pressure*, expressed in kilopascals—1 kPa = 75 mm Hg at 0°C = 10 mbar) increases with temperature (Fig. 1.5). However, air is usually not saturated and the vapour pressure is less than the saturated vapour pressure. For example, the vapour pressure at X is 10 mbar when the temperature (called *Dry-bulb Temperature*) is 18°C. Other ways of expressing the water vapour concentration at X are derived by drawing various lines from it to the saturated vapour pressure

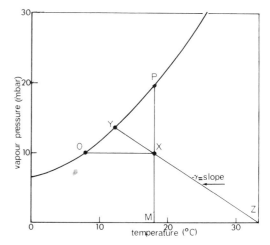

FIG. 1.5. Influence of temperature on the saturated vapour pressure of water vapour. The point X represents air at 18°C and 10 mbar vapour pressure. The line YXZ with a slope of $-\gamma$ gives the wet-bulb temperature from Y (12°C). The line OX gives the dew-point temperature from O (7.1°C). The line XP gives the saturated vapour pressure from P (20.6 mbar).

curve:

Dew-point Temperature is the temperature at which vapour pressure equals the saturated vapour pressure if air is cooled without gain or loss of water. Thus in Fig. 1.5, the temperature at O is the dew-point temperature of air (7.1°C) which has a vapour pressure of 10 mbar at 18°C.

Wet-bulb Temperature is the lowest temperature to which air can be cooled by evaporation of liquid moisture into air. It is the temperature to which a wet bulb falls in a wet- and dry-bulb psychrometer. In Fig. 1.5, Y is the wet-bulb temperature (12°C) and is found by drawing a line of slope $-\gamma$ through X (where γ is the psychrometic constant). If the wet bulb is ventilated with a pump or fan it reaches a lower temperature than one with natural ventilation. Thus a different psychrometric constant (slope) is used to calculate wet-bulb temperature with ventilated and non-ventilated psychrometers.

Saturation Deficit is the difference between the vapour pressure (e.g. at X) and the saturated vapour pressure at that dry-bulb temperature (P, 20.6 mbar in Fig. 1.5); that is, P − X (20.6 − 10 = 10.6 mbar). In other words, saturation deficit is an index of the drying power of the air; the higher the deficit the greater the evaporation rate. If air and leaf temperature are the same, saturation deficit is equivalent to the air/leaf vapour pressure difference $(e_l - e_a)$ and is directly related to the transpiration rate T:

$$T = \frac{e_l - e_a}{r_a + r_s} \qquad (1.2.2)$$

where r_a and r_s are, respectively, the boundary layer and stomatal resistances to water vapour transfer.

Relative Humidity is the ratio of the vapour pressure (X) to the saturated vapour pressure (P) at that dry-bulb temperature (Fig. 1.5), and is expressed as a percentage $(X/P \cdot 100\%)$. Relative humidity is mainly used to describe the moisture content of air as it influences human comfort. It has no direct influence on any plant biological processes and should therefore not be used. Instead, one or more of the other parameters described above are preferable. One common misuse in controlled environment studies is to hold relative humidity constant in order to keep evaporation rate constant while varying temperature experimentally. This results in saturation deficit, and hence evaporation rate, increasing with temperature. Saturation deficit, *not* relative humidity, should be kept the same at the different temperatures.

All these parameters which describe the water vapour content of air are interconvertible if the dry-bulb temperature is known. These inter-relationships are shown on a psychrometric chart (Fig. 1.6); for example, if wet- and dry-bulb temperatures are 10 and 20°C, respectively, relative humidity is 50%, humidity ratio is 7.5 g water kg^{-1} air, and vapour pressure is 8.5 mm Hg.

1.2.3.3. Methods of Measurement

Humidity sensors work on one of three principles: wet-bulb depression, relative humidity, or dewpoint temperature (Gaffney, 1978).

1.2.3.3(i). WET- AND DRY-BULB PSYCHROMETRY

A wet and dry psychrometer consists of two matched temperature sensors, one of which is covered with a wet muslin wick. Evaporation cools the wetted sensor to the wet-bulb temperature. Water vapour pressure (e) is calculated from the following formula:

$$e = e_s(T') - \gamma(T - T')$$

where T' and T are, respectively, wet- and dry-bulb temperatures, $e_s(T')$ is the saturated vapour pressure at wet-bulb temperature, and γ is the psychrometric constant, the value of which depends on whether the psychrometer is ventilated or non-ventilated. Values obtained with ventilated psychrometers such as Assman or whirling psychrometers are usually more accurate than the cheaper, non-ventilated types used in meteorological screens. Wet- and dry-bulb psychrometers are relatively inexpensive and simple. However, their accuracy is seriously impaired at relative humidities <20% and at temperatures approaching 0°C.

1.2.3.3(ii). RELATIVE HUMIDITY SENSORS

Relative humidity sensors vary from rather simple devices where relative humidity influences the mechanical properties of a material (e.g. the length of animal hair in simple hygrometers) to the more complex ones where it affects the electrical properties of sensors such as lithium chloride, sulphonated polystyrene or thin-film solid state semiconductors. Mechanical sensors are inexpensive but not very accurate, whereas the electrical sensors are more expensive but offer

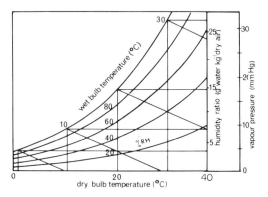

FIG. 1.6. Psychrometric chart illustrating the relationships between dry-bulb temperature and wet-bulb temperature, relative humidity (RH), humidity ratio and vapour pressure.

improved accuracy, ruggedness, compactness and remote electronic read out.

The lithium chloride sensor is the most common type of electrical sensor and it is relatively cheap. Lithium chloride is hygroscopic and the moisture content of the air determines how much water is absorbed, which in turn influences the AC resistance of the sensor. This type of sensor is susceptible to contamination by dust and other hygroscopic particles, and it suffers from hysteresis upon whether it is being wetted up or dried down. All electrical sensors are sensitive to changes in temperature, for which a correction must be made either electrically or by calculation.

Another simple and cheap method is based on the colour of cobalt chloride impregnated in paper, changing from blue at low relative humidity through a series of lilac shades to pink at high relative humidity. Such indicators are available commercially covering a range from 10 to 100% relative humidity.

An even simpler method which can also be used in small spaces is to measure the change in weight of a 10 mg piece of paper with a laboratory balance. The amount of water absorbed by the paper is proportional to the relative humidity which can be determined to an accuracy of 2% (Penman, 1955).

1.2.3.3(iii). Dew-point Hygrometry

There are basically two types of dew-point sensors: saturated salt, and condensation-hygrometer.

The saturated-salt types are widely used because of their low cost, simplicity and ruggedness. In addition, they are not affected by comtaminating ions. Their major limitation is slow response time and inability to measure below 10% relative humidity. The principle of operation is described by Gaffney (1978) and some of the references he quotes.

The condensation-type dew-point hygrometers operate over a wider range of dew-points and are faster, more accurate, and more reliable. However, they are more expensive and more complex than saturated-salt types. A surface in contact with the atmosphere to be measured is cooled until dew begins to form. This is detected optically or electrically, and cooling ceases. When the dew evaporates cooling recommences. The equilibrium temperature at which dew is just forming and evaporating is the dewpoint temperature. As long as the surface temperature-measuring system is calibrated, the instrument is fundamentally self-calibrating. Dewpoint temperatures must be corrected for changes in atmospheric pressure if they are converted into vapour pressure.

Bibliography and Further Reading

Campbell, C. S. (1977) *An Introduction to Environmental Biophysics.* (Springer-Verlag: New York.)

Gaffney, J. J. (1978) Humidity: basic principles and measurement techniques. *Hort. Science* **13**, 551–555.

Monteith, J. L. (1973) *Principles of Environmental Physics.* (Edward Arnold: London.)

Penman, H. L. (1955) *Humidity.* (Chapman & Hall: London.)

1.3. BIOMASS ACCUMULATION AND PRIMARY PRODUCTION

by L. L. TIESZEN

1.3.1. Introduction

The process of photosynthesis results in the conversion of light energy to chemical energy. This energy input by PRIMARY PRODUCERS is used to do *work* or is *stored*. This stored chemical potential is our main concern in this training course on bioproductivity and photosynthesis, since it is this BIOMASS which can be harvested for food, fuel, fibre or other practical uses. It is obviously of equal importance in natural ecological systems since this biomass provides the energy source and organic molecules for all other trophic levels.

The rate of energy input into a system as a result of photosynthesis is our measure of GROSS PRODUCTIVITY, P_g. Some of this fixed energy is used by plants for work processes requiring RESPIRATION, R. The NET PRODUCTIVITY, P_n, is the energy remaining in the system and therefore represents the rate of energy storage. These relationships are summarized:

$$P_g = P_n + R$$

or

$$P_n = P_g - R.$$

It is clear that, when $P_g = R$, there is no accumulation of stored energy (i.e. $P_n = 0$). Under these conditions the system is simply maintaining itself and the biomass, B, will not change. Systems of high net productivity are therefore obviously characterized by rapid increases in biomass.

It is obvious that we can estimate net productivity from gas-exchange measurements, if we know the respiratory losses from all compartments (i.e. roots, stems, flowers, etc., as well as leaves) for light and dark periods. P_g for a plant should equal the apparent photosynthetic rates for all photosynthetic structures plus the "dark" respiration rates of those structures. Net productivity, P_n, for the plant will be much less than the sum of apparent ("net") photosynthesis of leaves because of respiratory losses from other compartments. It can be calculated as the apparent photosynthesis of all productive tissues minus the respiration rates of all non-photosynthetic structures. If all these components are correctly accounted for, the *net* CO_2 incorporation for the plant should be equivalent to the net productivity, P_n, or the net increase (change) in biomass. We can assume that $1 \, g \, CO_2$ is equivalent to $0.648 \, g$ saccharide. Since plant biomass is mainly carbohydrate, this conversion is reasonably accurate for estimating biomass.

Gas-exchange studies are usually undertaken to study the mechanism controlling net production rather than to estimate net production. This is because it is usually easier to estimate net productivity by direct biomass harvests during a known time interval. Under ideal conditions:

$$\frac{\Delta B}{\Delta t} = P_g - R,$$

or

$$\frac{\Delta B}{\Delta t} = P_n,$$

or

$$P_n = \frac{\Delta B}{\Delta t}.$$

Under many conditions, however, this simple relationship will underestimate P_n because biomass is lost due to DEATH and DECOMPOSITION, L, or is utilized or HARVESTED, H, by herbivores. Thus, the most correct determination of P_n is:

$$P_n = \frac{\Delta B}{\Delta t} + \frac{L}{\Delta t} + \frac{H}{\Delta t}.$$

That is, net productivity, P_n, is equal to the change in biomass over a given time interval plus the biomass lost to decomposition and death, L, plus the biomass lost to herbivory, H, during the same interval.

Productivity is a rate function and is normally expressed on a unit land (soil) area basis. Common units are:

$$g \, m^{-2} \, d^{-1}$$

or

$$J \, m^{-2} \, d^{-1}.$$

The relationship between weight and energy can be established by determining the calorific value of the dry matter under study. This is readily accomplished with a bomb calorimeter.

Net productivity is sensitive to many factors both internal and external to the plant. It will be a function of the intercepting ability of the plant. Hence canopy density (LAI), leaf inclination, light absorbance, and allocation to synthesize new leaves are all important. In addition, we would expect P_n to be positively related to rates of apparent photosynthesis, if all other factors are equal. It is therefore clear that nutrition, water status, irradiance and many other environmental factors will influence P_n.

1.3.2. Net Productivity Studies in Natural Ecosystems

Although conceptually simple, the accurate measurement of net primary production in nature is difficult. The term refers to biomass changes in the entire plant. Yet it is nearly impossible to delimit below-ground compartments and to measure realistic changes in these compartments. Thus, most studies are restricted to *Net Above-ground Primary Production* (NAPP) and changes in this compartment only are measured. When only this compartment is measured, there is the inherent danger that changes in B (ΔB) are due, in part, to translocation (into or out of the above-ground compartment) and not only due to changes in net photosynthesis as is often assumed.

In many temperate-zone systems the time interval over which NAPP is measured begins with the onset of growth in spring and terminates at the "peak" of the season. It has been shown numerous times that this grossly underestimates NAPP because (1) all species do not "peak" at the same time *and* (2) there is a constant turnover of leaves, i.e. grass leaves die and begin to decompose continuously. These two problems are enhanced in tropical areas where distinct seasonality may be rare and where leaf turnover may be very rapid. The following describes a technique which minimizes these problems.

Net above-ground primary production is usually expressed by the following symbols:

B_1 = biomass of plant community at time t_1,
B_2 = biomass of plant community at time t_2,
ΔB = changes in biomass during the period $t_2 - t_1$ or $B = B_2 - B_1$,
L = plant loss by death and shedding during $t_2 - t_1$,
H = plant loss by consumer organisms during $t_2 - t_1$,
P_n = net production during $t_2 - t_1$.

If ΔB, L and H are correctly estimated, then

$$P_n = \frac{\Delta B}{\Delta t} + \frac{L}{\Delta t} + \frac{H}{\Delta t}.$$

The difficulty in natural systems is that the compartment, L, loses weight (and energy) rapidly because decomposition can be very effective and rapid. Therefore, a correct estimate of P_n must take into account changes in the size of this compartment *and* the losses from it by decomposition.

We can calculate the disappearance of dead material by decomposition using a paired plots or litter bag system by:

$$r_i = \frac{\log_e(W_0/W_1)}{t_1 - t_0}$$

where r_i = disappearance rate of dead plant parts in g g^{-1} day^{-1},
W_0 = mean amount of dead material collected from the quadrats at time t_0 in g m^{-2},
W_1 = the amount of dead material collected from matching quadrat at time t_1 minus h (the amount of dead material added to it during $t_1 - t_0$ in g m^{-2}).
$t_1 - t_0$ = time interval in days.

Establish a statistically significant number of matched quadrats. Remove all the dead material from one half of the quadrats. At the end of the time interval in question, sample both sets of quadrats and separate into live and dead compartments. Now that we know the standing crop of the live (B) and dead (A) compartments at the beginning and end of a time interval of t_1 days, *and* if we know r_i, and if we excluded all herbivores, we can calculate NAPP.

t_i = time interval in days,
A_1 = g m^{-2} dead at start of interval,
A_2 = g m^{-2} dead at end of interval,
B_1 = g m^{-2} live at start of interval,
B_2 = g m^{-2} live at end of interval,
r_i = instantaneous rate of disappearance of dead (g g^{-1} d^{-1}).

The change in live (ΔB) is:

$$\Delta B_i = B_2 - B_1.$$

The change in dead (ΔA) is:

$$\Delta A_i = A_2 - A_1.$$

Now, the amount of dead disappearing to decomposition we will call X_i and calculate by:

$$X_i = \frac{A_2 + A_1}{2} r_i t_i.$$

Therefore, the total amount of dead material (L_i) added in the interval is:

$$L_i = X_i + \Delta A_i.$$

We have now accounted for the second term (L_i) in our calculation of NAPP. The third term H_i is very difficult to account for totally, and methods to exclude the herbivores or account for their effects are clearly species specific and are beyond the scope of this topic.

1.3.3. Leaf-area Measurement

Total leaf area is often a useful measurement in bioproductivity studies. There are various instrumental methods available for leaf-area measurement. For specific plants, leaf area can often be derived from length and breadth measurements.

$$\text{Leaf area} = A = kLB,$$

A = area,
L = leaf length,
B = breadth at mid point,
k = constant determined for species under investigation.

The Lambda leaf area meter Licor (Lincoln, Nebraska, U.S.A.) is a very useful, portable and non-destructive measuring unit. It consists of a hand-held *sensor* and a *readout unit* which displays area in cm^2. The actual sensor consists of two parts. One is a row of lights spaced one every millimetre, and a row of light detectors positioned just opposite each light (one each mm, one specific for each light). The leaf is placed between the row of lights and the row of detectors. Leaf area, then, is detected blocked light. With no leaf in the sensor, no light is blocked, all the detectors receive light, and no area is recorded. With a stationary leaf positioned in the sensor, the number of blocked lights is the area of the leaf (mm^2) for that mm of leaf length. Holes in the leaf present no problem, for they will allow light to pass and so will not contribute to the area. The other main part of the unit is the length *encoding cord*. This is pulled out at the same rate as the leaf is pulled through the sensor. It causes the sensor to determine the number of blocked lights once for each mm the cord is pulled (one flying spot scan for each mm length). The result is that the readout unit records the "width" for each mm of leaf length pulled through the sensor. Then it

sums these areas, converts from mm^2 to cm^2, and displays the result. The unit is portable, operating on rechargeable batteries, or from 110 or 240 V.

To actually measure a leaf, first place the petiole between the light row and detect row. Then pull out the length encoding cord and hold it first to the petiole (or wrap around the stem). While positioning the leaf and cord, some area may be recorded, so when ready to go, press the zeroing button on the hand-held sensor. Release this, and pull the leaf through the sensor. The rate of leaf movement through the sensor need not be constant. When the leaf is completely through, read the area and let the cord *slowly* rewind itself back into the sensor. The leaf does not have to be removed from the plant to be measured.

There are several sources for errors. Curling of leaf edges may give some errors. Also, if the petiole is in the detecting row at the start, of course petiole area will be added to leaf area (although this will be small). If the length cord is not pulled out straight, or if it is not pulled out at the same rate as the leaf, this will again result in errors. However, with high operator proficiency, precision of $1 \, mm^2$ can be obtained.

Bibliography and Further Reading

CHAPMAN, S. B. (1976) *Methods in Plant Ecology.* (Blackwell: Oxford.)

GOLLEY, F. B. (1961) Energy values for ecological materials. *Ecology* **42**, 581–584.

MILNER, C. and HUGHES, R. E. (1968) *Methods for the Measurement of Primary Production of Grassland.* IBP Handbook. (Blackwell: Oxford.)

ŠESTÁK, Z., ČATSKÝ, J. and JARVIS, P. G. (1971) *Plant Photosynthetic Production. Manual of Methods.* (Dr. W. Junk: The Hague.)

WIEGERT, R. G. and EVANS, F. C. (1964) Primary production and the disappearance of dead vegetation on an old field in southeastern Michigan. *Ecology* **45**, 49–63.

Appendix 1.3. Experimental Work: Biomass Accumulation and Primary Production.

Experiment

An experiment which was used in the training courses and can be completed by a group of students within two half-day sessions is the measurement of dry weight increase in a stand of garden beans (*Phaseolus vulgaris*).

Objectives

The objectives were to illustrate a simple method of measuring productivity and some factors which will affect the magnitude of P_n. If parallel gas analysis measurements are made (Appendix 1.5), then P_n measured by this harvest method may be compared with P_n estimated from gas analysis data. The exercise may also illustrate simple experimental design and statistical considerations in productivity studies.

Equipment Required

A balance accurate to 0.1 g (plus calibration weights, if available).

A drying oven, preferably with forced circulation.

A leaf area meter or materials for measuring leaf area, i.e. graph paper or photosensitive paper.

Graph paper and calculators.

Meteorological data for the preceding 10 days.

Any available data on productivity of the same species for other sites and conditions.

Plant Material

Sow an excess of seeds of *P. vulgaris* each in 8 cm pots, and arrange these into blocks of equal size. About 3 weeks after germination and 10 days before the student practical, make one harvest of half of the plants. Provide the students with this data.

These plants should be grown without check until the first harvest—that is, in a good compost such as John Innes No. 2, at about 25°C and in good light. After the first harvest, withdraw irrigation from one-half of the remaining pots to provide a drought treatment.

Procedure

The students are provided with the following information.

In this experiment you will study the effect of water stress on the net productivity of *P. vulgaris*. You will make the final harvest of two treatments, one supplied with ample water and the other

droughted. A previous harvest of leaves, stems and roots was made 10 days prior to your harvest. You are to calculate P_n during this 10-day interval.

1. Select, by a randomized design, five plants for each treatment. Determine the mean land area occupied by each pot.
2. Count the number of leaves in the five strata of the canopy. Randomly identify fifteen leaves per stratum and measure leaf area.
3. Estimate the mean leaf inclination for each stratum by measuring twenty to thirty leaves per stratum.
4. Harvest all leaves (without petioles) per pot, and all stems per pot.
5. Separate soil from all roots with the sieve and washing device supplied.
6. Identify each pot and plant compartment and place in a drying oven (75°C) until a constant weight is obtained—usually about 24 hours.
7. Calculate P_n according to the previously described equations, assuming L and H to be zero.

The results obtained can be related to meteorological data for the period. This is particularly instructive if you have data for other crops and conditions. The data may also be used for the calculation of growth analysis parameters using the classical procedures (Section 1.4).

1.4. PLANT-GROWTH ANALYSIS

by C. L. BEADLE

1.4.1. Introduction

If we wish to measure the bioproductivity of a natural ecosystem or agricultural crop, the component of immediate interest is the total yield or net primary production. As pointed out in Section 1.3, it is usually necessary to restrict our interest to above ground parts and in agricultural crops only the economic yield, e.g. the grain of cereals may be of importance. This is commonly expressed as the Harvest Index:

$$\text{Harvest Index} = \frac{\text{Economic Yield}}{\text{Biological Yield}} \times 100. \quad (1.4.1)$$

These figures tell us little about the changes which are occurring during the growth of the crop however. Similarly, photosynthetic gas-exchange analysis (Section 1.5) can be used to describe the responses of carbon fixation to changes in the environment and may eventually be used to measure total yield, but no information is gained about the partitioning of dry matter into new leaf area which can have a considerable influence on production.

In 1919 Blackman defined production in terms of a compound interest law ("If the rate of assimilation per unit area of leaf surface and the rate of respiration remain constant, and the size of the leaf system bears a constant relation to the dry weight of the whole plant, then the rate of production of new material, as measured by the dry weight will follow the compound interest law") and over the last 60 years this has been the basis for the development of the techniques of plant-growth analysis. These techniques can answer some of the questions posed above and serve as an intermediate experimental tool between those of gas-exchange and net primary production.

1.4.2. Basic Principles

Two types of measurement only are needed for growth analysis:

(i) The plant weight. This is usually the oven dry weight (kg) but can be the organic matter or energy content.

(ii) The size of the assimilatory system. This is usually the leaf area (m^2) but can be the leaf protein or chlorophyll content.

This primary data of growth analysis can be made on individual plants or derived from whole canopies, though the destructive nature of the techniques requires the use of homogeneous sets of plants or plots. In its simplest form therefore plant-growth analysis requires little more than a balance, photosensitive paper and a calculator for quite detailed studies of quantitative aspects of dry-matter production. Accurate measurements are essential, however, to avoid unnecessary variation during sampling.

Terminology

This quantitative description of growth is based on several terms. As these have been developed over a number of years these terms and the symbols which represent them have evolved accordingly. The terminology used here will follow that of Hunt (1978) which refers to the components in terms of single letters which readily lend themselves to mathematical notation.

1.4.3. Components of Classical Growth Analysis

Relative Growth Rate, R

The basic component of growth analysis, which arose from the work of Blackman (1919) referred to above, is the relative growth rate, R, of the plant or crop. This is defined at any instant in time t, as the increase of material per unit of material present and is the only component of growth analysis which does not require knowledge of the size of the assimilatory system. Thus,

$$R = \frac{1}{W}\frac{dW}{dt} = d\frac{\ln W}{dt}. \qquad (1.4.2)$$

The relative growth rate is therefore the in-

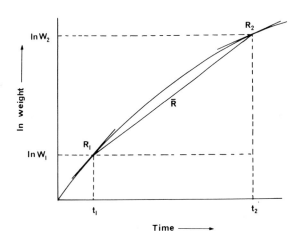

FIG. 1.7. The derivation of instantaneous and mean relative growth rates from a plot of ln W against t (redrawn from Hunt, 1978).

stantaneous slope of the ln W versus time curve (kg/day, Fig. 1.7).

In practice, the mean relative growth rate, $\bar{R}$, is measured over a discrete time interval, t_1 to t_2, which is normally no less than 1 day. As long as W varies continuously from t_1 to t_2, $\bar{R}$ is defined as

$$\bar{R} = \frac{1}{t_2 - t_1} \int_{W_1}^{W_2} d(\ln W) = \frac{\ln W_2 - \ln W_1}{t_2 - t_1}.$$

(1.4.3)

The relationship between R and $\bar{R}$ is shown in Fig. 1.7. It should be noted that in the special case of exponential growth, i.e. the form $W = W_0 e^{Rt}$, R is constant and equal to $\bar{R}$ (N.B. W_0 is the initial weight when $t = 0$).

The relative growth rate serves as a fundamental measure of dry-matter production and can be used to compare the performance of species or the effects of treatments under strictly defined conditions. As with measurements of photosynthesis, however, R tells us little about the causal factors which determine that performance. These are encompassed by the concepts of Unit Leaf Rate and Leaf Area Ratio.

Unit Leaf Rate, E

The unit leaf rate, E, of a plant or crop at any instant in time t is defined as the increase of plant material per unit of assimilatory material per unit of time:

$$E = \frac{1}{A} \frac{dW}{dT}.$$

(1.4.4)

The term net assimilation rate is often used interchangeably with unit leaf rate but the latter is preferred (Evans, 1972). They measure the net gain in dry weight of the plant per unit leaf area (kg/m^2) and differ from photosynthetic rate which measures the net gain of carbon only during the light period.

The mean unit leaf rate, $\bar{E}$, between t_1 and t_2 is given by

$$\bar{E} = \frac{1}{t_2 - t_1} \int_{t_1}^{t_2} \frac{1}{A} \frac{dW}{dt} \, dt.$$

(1.4.5)

As Radford (1967) has pointed out this relation-

ship cannot be integrated as above unless

(i) the relationship between A and W is known,
(ii) the relationships between A and t, and W and t are known.

If one assumes, however, that

(iii) over the period t_1 to t_2, A and W are linearly related and
(iv) A and W are not discontinuous functions of time equation (1.4.5) can be integrated such that

$$\bar{E} = \frac{W_2 - W_1}{A_2 - A_1} \frac{\ln A_2 - \ln A_1}{t_2 - t_1}.$$

(1.4.6)

Radford (1967) lists further formulae for the calculation of $\bar{E}$ where the relationship between W and A is known but is not linear.

Leaf Area Ratio, F

The leaf area ratio, F, of a plant or crop at any instant in time, t, is the ratio of the assimilatory material per unit of plant material present:

$$F = \frac{A}{W}.$$

(1.4.7)

Using the same assumptions (iii) and (iv) above the mean leaf area ratio, $\bar{F}$ is given by

$$\bar{F} = \frac{A_2 - A_1}{W_2 - W_1} \frac{(\ln W_2 - \ln W_1)}{(\ln A_2 - \ln A_1)}.$$

(1.4.8)

From a combination of equations (1.4.4) and (1.4.7) it can be seen that

$$R = E \times F = \frac{1}{W} \frac{dW}{dT} = \frac{1}{A} \frac{dW}{dT} \times \frac{A}{W}$$

(1.4.9)

The relative growth rate therefore consists of two components which measure the efficiency of the plant or crop as a producer of dry weight (E) and a producer of leaf area (F). Although equation (1.4.9) is true at any instant, the same relationship with respect to the mean values does not hold (i.e. $\bar{R} = \bar{E} \times \bar{F}$) as except under very special circumstances the relationship between A and t, W and t, and A/W and t are not linear and rarely take the same form (Radford, 1967).

Specific Leaf Area and Leaf Weight Ratio

The leaf area ratio can be redefined as

$$F = \frac{L_w}{W} \times \frac{A}{L_w}. \qquad (1.4.10)$$

where L_w is the dry weight of the leaves. The two components, L_w/W and A/L_w are called the leaf weight ratio (LWR) and specific leaf area (SLA) respectively. SLA defines leaf area ratio in terms of leaf density (i.e. m^2/kg) and its reciprocal, LWR (kg/m^2) is a measure of the leafiness of the plant on a weight basis.

Leaf Area Index, L

If we specifically wish to consider the productivity of crops or natural ecosystems, it is convenient to express their performance per unit land area. The leaf area ratio is therefore inappropriate and a second term, the leaf area index, L (Watson, 1947), or leaf area per unit land area is used. This can be written as

$$L = \frac{L_a}{P} \qquad (1.4.11)$$

where L_a is the functional leaf area of the crop canopy standing on ground area P. As both L_a and P are normally measured as areas (m^2), L is dimensionless. L_a can refer to the total surface area of the leaves, i.e. both the upper and lower surface, the area of a single surface, or the projected area where leaves are other than flat, e.g. conifers. Care should be taken when making comparisons that L is measured on the same basis.

Crop Growth Rate, C

The leaf-area index can be used to calculate the instantaneous crop growth rate (at any time t) which serves as a simple index of agricultural productivity. This is defined as

$$C = E \times L = \frac{1}{P}\frac{dW}{dT} = \frac{1}{L_a}\frac{dW}{dT} \times \frac{L_a}{P} \qquad (1.4.12)$$

and is expressed in terms of the weight per unit area and time ($kg/m^2 \times s$). C is therefore conceptually similar to R and for the same reasons, $\bar{C}$ is rarely equal to the product, $\bar{E} \times \bar{L}$. The elements of crop-growth rate can be used to assess the relative importance of variation in L or E as determinants of crop-growth rate. In general it is concluded that C is more sensitive to L than E (e.g. Watson, 1952; Potter and Jones, 1977).

Leaf Area Duration, D

The final component of growth analysis which will be considered here is the leaf area duration, D, a measure of the persistence of the assimilatory surface. There is no instantaneous value and it is normally calculated from the relationship between L and time (Fig. 1.8) though a similar measurement can be made on individual plants. As above for $\bar{E}$ and $\bar{F}$, D can be determined by integral calculus. An alternative method, although not mathematically correct, is to measure the area of a trapezium under that part of the curve which is of interest. Thus,

$$D = \frac{(L_1 - L_2)}{2}(t_2 - t_1); \qquad (1.4.13)$$

as this is the product of a dimensionless unit and time, the units of D are time and are usually expressed in days. The duration of the assimilatory area is considered to be of similar importance to the area itself (Watson, 1956). An approximate estimate of the total yield ($kg\ m^{-2}$) can now be defined from the product of D and

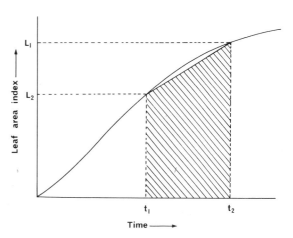

FIG. 1.8. Leaf-area duration (shaded area) determined from a plot of leaf-area index against time.

$\bar{E}$, the mean unit leaf rate. Thus,

$$\text{Yield} = D \times \bar{E}.$$

The estimate is approximate, as in practice $\bar{E}$ cannot be determined accurately (see above). The equation suggests, however, that maximum rates of dry-matter production will occur when maximum L coincides with maximum $\bar{E}$.

1.4.4. Investigations

Limitations

The techniques of growth analysis are most suited to following the effects of long-lasting treatments. They are particularly useful for studies of dry-matter production in relation to varietal differences of crop plants or agronomic practice, e.g. mineral nutrition, spacing and irrigation. The limitations in accuracy which the technique offers precludes its use in the investigation of subtle changes in climate which occur in the field as the variance attributable to sampling will in this instance exceed that of the factor investigated. This does not exclude the study of the effects of environmental factors in controlled environments by plant-growth analysis.

Experimental Procedure

Statistically homogeneous sets of plants or pots are required for each sequential harvest so that dry-weight increments are not obscured by unwanted variation. Plants should be selected on the basis of size at the start of the experiment and paired or grouped in such a way that each set to be harvested contains plants with similar morphology and growth rate (e.g. Goodall, 1945; Evans and Hughes, 1961). Experimental plots of similar uniformity should be selected at the start of field crop experiments, and similar criteria should be applied to natural vegetation if possible.

The number of replicates at each harvest and the harvest interval should be adjusted to the growth rate such that the effects of the treatments are not obscured by sampling variation. In practice, it is recommended that this interval

should be timed according to changes in A and that $A_2/A_1 = 2$ for satisfactory measurements of $\bar{E}$ (see Květ et al., 1971).

The statistical procedures applied to growth analysis are those which are suitable for any experiment where statistically homogeneous sets of data are collected (e.g. see Snedecor and Cochran (1972) for the proper application of statistics in agriculture).

Functional Growth Analysis

General

The classical form of growth analysis described above has been joined but not superseded by a second approach called "functional growth analysis" (see Hunt, 1978). This has arisen partially from the availability of computer-based facilities for fitting functions to data and also from the limitations of classical growth analysis, viz. the assumptions discussed above. This functional or dynamic approach was proposed by Radford (1967) and is based on more frequent and smaller harvests (1 to 3 days) when the grouping or pairing of plants can be avoided. The data is then used to accurately and adequately describe the relationships between A and t, and W and t which are fitted with appropriate functions. These are usually polynomial in form. The major advantage of this approach is that information for the whole period of interest is contained in two equations and further functions can be developed to calculate the instantaneous values of R, E and F over the same period. A critical summary of the development and use of functional growth analysis is given by Hughes and Freeman (1967), Vernon and Alison (1963), Hunt (1978, 1979).

The Richards Function

Functional growth analysis has also made use of the Richards (1959) growth function as well as polynomial equations. This function takes the form

$$W = a(1 \pm e^{(b-kt)^{-1/n}}) \qquad (1.4.14)$$

where *a*, *b*, *k* and *n* are constants. The development of this equation has been considered by Causton *et al.* (1978) who show that *R* is a function of a declining linear function of *n*, and that the size of *n* determines the shape of the relationship between *R* and *W*. The function therefore has considerable flexibility and Venus and Causton (1979) contend that it provides a more biologically meaningful fit than a polynomial equation when fitted to data collected over several days.

Bibliography and Further Reading

BLACKMAN, V. H. (1919) The compound interest law and plant growth. *Ann. Bot.* **33**, 353–360.

CAUSTON, D. R., ELIAS, C. O. and HADLEY, P. (1978) Biometrical studies of plant growth. I. The Richards function, and its application in analysing the effects of temperature on leaf growth. *Plant Cell Env.* **1**, 163–184.

EVANS, G. C. (1972) *The Quantitative Analysis of Plant Growth.* (Blackwell Scientific, Publ.) 734 pp.

EVANS, G. C. and HUGHES, A. P. (1961) Plant growth and the aerial environment. Effect of artificial shading on *Impatiens parviflora*. *New Phyt.* **60**, 150–180.

GOODALL, D. W. (1945) The distribution of dry wt change in young tomato plants. Dry weight changes of the various organs. *Ann. Bot.* N.S. **9**, 101–139.

HUGHES, A. P. and FREEMAN, P. R. (1967) Growth analysis using frequent small harvests. *J. Appl. Ecol.* **4**, 553–560.

HUNT, R. (1978) *Plant Growth Analysis. Studies in Biology* No. 96. (Edward Arnold, Publ.) 67 pp.

HUNT, R. (1979) Plant growth analysis: The rationale behind the use of the fitted function. *Ann. Bot.* N.S. **43**, 245–249.

KVĚT, J., ONDOK, J. P., NEČAS, J. and JARVIS, P. G. (1971) Methods of growth analysis. In: *Plant Photosynthetic Production* (Eds. Z. ŠESTÁK, J. ČATSKÝ and P. G. JARVIS) (Dr. W. Junk N.V., publ.), pp. 343–391.

POTTER, J. R. and JONES, J. W. (1977) Leaf area partitioning as an important factor in growth. *Plant Physiol.* **59**, 10–14.

RADFORD, D. J. (1967) Growth analysis formulae—their use and abuse. *Crop. Sci.* **7**, 171–175.

RICHARDS, F. J. (1959) A flexible growth function for empirical use. *J. Exp. Bot.* **10**, 290–300.

SNEDECOR, G. W. and COCHRAN, W. G. (1972) *Statistical Methods Applied to Experiments in Agriculture and Biology.* (Iowa State College, Publ.) 593 pp.

VENUS, J. C. and CAUSTON, D. R. (1979) Plant growth analysis: the use of the Richards function as an alternative to polynomial exponentials. *Ann. Bot.* N.S. **43**, 623–632.

VERNON, A. J. and ALLISON, J. C. S. (1963). A method of calculating net assimilation rate. *Nature* **200**, 814.

WATSON, D. J. (1947) Comparative physiological studies on the growth of field crops. I. Variation in net assimilation rate and leaf area between species and varieties and within and between years. *Ann. Bot.* N.S. **11**, 41–76.

WATSON, D. J. (1952) The physiological basis of variation in yield. *Adv. Agron.* **4**, 101–145.

WATSON, D. J. (1956) Leaf growth in relation to crop yield. In: *The Growth of Leaves* (Ed. F. L. MILTHORPE) (Butterworths, Publ.), pp. 178–191.

WILLIAMS, R. F. (1975) The quantitative description of growth. In: *The Shoot Apex and Leaf Growth.* (Cambridge University, Publ.), 256 pp.

1.5. MEASUREMENT OF PHOTOSYNTHETIC GAS EXCHANGE

by S. P. LONG

1.5.1. Introduction

Photosynthetic gas exchange in this context refers to the fluxes, or flows, of gas between the plant and the atmosphere produced through photosynthesis. The gas most commonly measured is CO_2 since it provides a direct measurement of productivity. Simultaneous measurements of water vapour and also O_2 are valuable in providing further information on limitations to photosynthesis in intact tissues. Modern techniques of determining CO_2 allow the detection of very small changes in concentration, thus sensitive determination of rates of CO_2 uptake by small areas, e.g. small leaves or even segments of leaves, is possible. Such techniques can be used to study the photosynthetic contribution of individual leaves (at different stages of development or in different conditions) to productivity in either field or controlled environments. Normally the material is enclosed in a chamber and CO_2 fluxes determined from measured CO_2 concentration changes in the chamber atmosphere. One alternative to the use of chambers is to measure CO_2 concentration profiles above large stands of vegetation and calculate CO_2 fluxes from

meteorological considerations; see Biscoe *et al.*
(1975) and Monteith (1975) for theory and
examples. Gas-exchange studies conveniently
fall into two categories: field studies (Chapter 1.6)
and laboratory or controlled environment studies which will be considered here. CO_2 assimilation has been measured by a wide range of
techniques, the most common being $^{14}CO_2$,
labelling, conductivity and IR spectroscopic
analysis. The latter, infra-red gas analysis of
CO_2, is the most widely used method in present
studies of leaf and whole-plant photosynthesis.

1.5.2. Infra-red Gas Analysis

1.5.2.1. Principle

Infra-red gas analysis is by far the most popular method of determining photosynthetic and
respiratory CO_2 exchange in plants. This popu-
larity stems from the reliability, accuracy and
simplicity of this technique when compared to
other available methods.

Heteratomic gas molecules typically absorb
radiation at specific infra-red wavebands; each
gas having a characteristic absorption spectrum.
On the other hand, gas molecules consisting of
two identical atoms (e.g. O_2, N_2) do not absorb
infra-red radiation (IR), and thus do not
interfere with determination of the concentration of heteratomic molecules (e.g. Banwell,
1966). Infra-red gas analysis has been used for
the measurement of a wide range of heteratomic
gas molecules, including CO_2, H_2O, NH_3, CO,
SO_2, N_2O, NO and gaseous hydrocarbons (Janáč
et al., 1971). Thus, infra-red gas analysis can be
used not only for the accurate determination of
CO_2 concentration but also H_2O vapour in transpiration studies and CO, SO_2, NO, etc., in
studies of atmospheric pollution.

The major absorption band of CO_2 is at $\lambda =
4.25\ \mu$m with secondary peaks at $\lambda = 2.66, 2.77$
and $14.99\ \mu$m. The only heteratomic gas normally present in air with an absorption spectrum
overlapping with that of CO_2 is water vapour
(both molecules absorb IR in the $2.7\ \mu$m region)
(Janáč *et al.*, 1971). Since water vapour is usually present in air at much higher concentrations
than CO_2 this interference does present a

significant problem. This is overcome either by
drying the air that is to be examined or by
filtering out all radiation at the wavelengths
where absorption by the two gases coincides.

1.5.2.2. Construction

The infra-red gas analyser consists of three
basic parts, the IR source, the sample chambers
and the detector. Figure 1.9 illustrates the construction of an IRGA with detector absorption
chambers in parallel—this is the most common
construction of commercially available instruments at the present time.

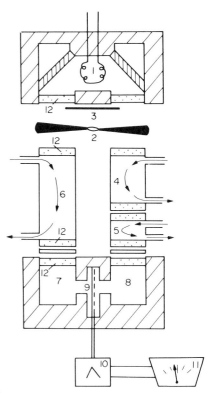

FIG. 1.9. Diagram of construction of an
IRGA with sample tubes in parallel, and
a split analysis tube. 1 = Nichrome
spiral IR source; 2 = Rotational shutter;
3 = Zero shutter; 4 and 5 = Analysis
tube (long and short cell, respectively);
6 = Reference tube; 7 and 8 = Absorption chambers; 9 = Microphone
detector; 10 = Amplifier and rectifier;
11 = Output meter; 12 = IR transmitting
windows.

The IR source (1) is a nichrome spiral which is heated to about 600–800°C (dull red glow) through a low-voltage circuit. The beam is split by reflectors so that approximately equal amounts of radiation are passed through both the reference and analysis sample tubes (4–6); the amounts of radiation being precisely balanced by the zero shutter (3). The tubes have highly reflective linings, typically gold coated, and are capped by IR transmitting "windows" (12). In the absolute mode of operation the reference tube is filled with nitrogen or CO_2-free air and the sample gas is passed through the analysis tubes. IR passes through the reference tube unaffected but is reduced in the analysis tubes due to absorption by CO_2 molecules, thus the amount of IR in the CO_2 absorption bands, reaching the detector is reduced in the analysis tube. The detector works on the principle of positive filtration, that is it absorbs IR in the CO_2 absorption bands. This is achieved by filling both absorption chambers with CO_2. Thus, the more CO_2 in the analysis tubes the greater the ratio of IR radiation in the CO_2 absorption bands reaching the reference chamber to that reaching the analysis chamber of the detector. The two chambers of the detector are separated by a diaphragm which forms one electrode of a diaphragm condenser (9). The paths of radiation are interrupted by a rotational shutter (2) causing periodical pressure charges in the detector with simultaneous vibration of the membrane. The amplitude of vibration is determined by the pressure difference between the two chambers which in turn is determined by the CO_2 concentration difference between analysis and reference tubes. Change in the amplitude of vibration of the membrane produces a change in the condenser capacity which is inversely proportional to a voltage change across the condenser. This change is amplified (10), rectified and measured by the output meter (11).

1.5.2.3. Calibration

Although the construction of the IRGA described allows highly sensitive and continuous monitoring of CO_2 concentration, the cost of these benefits is the lack of long-term stability of output calibrations. For any serious work it is essential to calibrate the analyser daily. The bare minimum requirement for reliable calibration is a source of CO_2 free air and a source of air containing a precisely known concentration of CO_2, in the range to be analysed and contained preferably in an aluminium cylinder (this should not adsorb CO_2 onto its walls as a steel cylinder will).

Absolute calibration. Where the analyser will be used to determine the exact CO_2 concentration of an air sample the analyser is calibrated in absolute mode, i.e. the sample is compared to CO_2-free air. To calibrate the analyser, CO_2-free air is passed through both the analysis and reference tubes. Zero is set on the output meter by adjusting the zero shutter. Air samples of known CO_2 concentrations are then passed through the analysis tubes starting with the highest concentration. The output meter deflection can be adjusted by the amplifier gain; however, such alterations will require the zero setting to be rechecked.

Differential calibration. Where the analyser will be used to determine a change in CO_2 concentration, for example the difference in CO_2 concentration in an air stream before and after it has passed over a leaf, the analyser is calibrated in differential mode. In this mode it is possible to detect very small changes in CO_2 concentration, down to $100 \, \mu g \, m^{-3}$ with some models. Precise calibration requires that the analysis and reference tubes are filled with air of known, but only slightly different, CO_2 concentrations. In practice this can only be achieved by a small dilution of the calibration gas using two precision gas-mixing pumps— these can each cost as much as an IRGA. However, an equally accurate differential calibration is possible, at no extra expense, on models in which the analysis tube is split—as illustrated in Fig. 1.9 (Parkinson and Legg, 1971). Here the analysis tube is split into two lengths, for example a long cell representing 95% (4) of the path length and a short cell representing the remaining 5% (5). To zero the instrument the air stream to be used in the experiment, and of previously determined absolute CO_2 concentration, is passed through both the reference and analysis tubes. A centre zero is then set on the output meter. At atmospheric concentrations of CO_2, the amount of IR absorbed in the analysis tube is only a small fraction of the total IR in the CO_2-absorbing bands which passes through the tube, thus removal of CO_2 from the short cell (by passing through CO_2-free air) is optically equivalent to reducing the CO_2 concentration throughout the length of the analysis tube by 5%. The change in the output meter produced by this operation represents the change that a 5% depletion in CO_2 concentration would produce, and thus provides the means of calibration.

1.5.3. Gas-analysis Systems

Infra-red gas analysers (IRGA) measure CO_2 concentrations. The rate of CO_2 uptake by a leaf can be measured by enclosing the leaf in a chamber, passing air over the leaf and measuring the change in CO_2 concentration with an IRGA. Such gas-analysis systems typically consist of three basic parts: an air-conditioning and supply unit, a leaf chamber and an IRGA. Three basic configurations of this equipment have been used and these are considered below (1.5.3.1–1.5.3.3).

1.5.3.1. Closed Systems

Closed systems (Fig. 1.10) are the simplest of the three and the most suitable for a laboratory working on a small budget and with little expertise in gas-analysis techniques. This type of system also makes the least demand on the sensitivity of your IRGA. In a closed system air is drawn from a chamber enclosing the leaf or plant into the analysis tubes of the IRGA, which is calibrated in absolute mode. The air is then recycled from the IRGA back to the chamber. Thus, no air will leave the system and no air will enter it from outside. If the leaf enclosed in the chamber is photosynthesizing then the CO_2 concentration in the system will decline, and continue to decline until the CO_2 compensation point of photosynthesis (Γ) is reached. Rate of photosynthetic CO_2 assimilation (F_{CO_2}) can be calculated from equation (1.5.1)

$$F_{CO_2} = \frac{\Delta C_a V}{tA} \qquad (1.5.1)$$

where ΔC_a = change in CO_2 concentration over time interval of length t (mg m^{-3}),
V = volume of system (m^3),
t = length of interval over which CO_2 concentration changes are recorded (s),
A = leaf area (m^2).

Thus to determine F_{CO_2} in a closed system the only requirements are: that the IRGA is calibrated accurately in absolute mode and that V and A are accurately determined. The major disadvantage of this type of system is that since the CO_2 concentration of the system will be changing, F_{CO_2} cannot reach steady state.

1.5.3.2. Semi-closed Systems

Semi-closed systems (Fig. 1.10) are a variation on the closed which allow C_a to remain constant and thus allow a steady-state F_{CO_2} to be attained. In this system the IRGA is used as a null-point instrument which controls through an electronic circuit a supply of CO_2 into the system. When CO_2 is removed by the photosynthesizing leaf a decrease in C_a sensed by the analyser switches on a supply of pure CO_2 to the system until the original C_a has been re-established.

$$F_{CO_2} = \frac{V \cdot 273 \cdot 44}{t \cdot A \cdot (273 + T_a) \cdot (22.4 \times 10^{-3})} \qquad (1.5.2)$$

where V = the volume of CO_2 added to the system over the time interval of the measurement (t),
t = length of interval over which CO_2 concentration changes are recorded (s),
A = leaf area (m^2),
T_a = air temperature.

To determine F_{CO_2} in a semi-closed system the requirements are: that the IRGA is calibrated in absolute mode, that A is accurately determined and that V the volume of CO_2 added to the system is known with absolute precision. To allow a truly steady-state value of F_{CO_2} to be attained the pulses of CO_2 addition should be as small and as frequent as possible: concentration fluctuations within the system should not exceed 10 mg m^{-3}. Thus this type of system

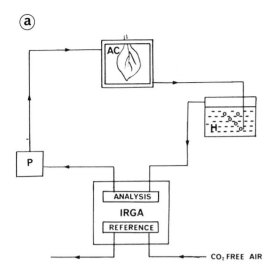

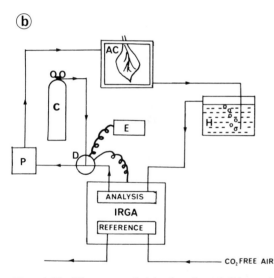

FIG. 1.10. Diagrams of (a) closed and (b) semi-closed gas-analysis systems. AC = Assimilation chamber; C = Cylinder of CO_2; D = Dosing cock; E = Electronic dose counter; H = Humidifier; IRGA = Infra-red gas analyser; P = Pump. Arrows indicate direction of air flow.

makes greater demand on the IRGA in that it has to accurately detect small changes in C_a and its output should be stable over a long period. In such a system care should be taken that a significant change in O_2 concentration does not occur as this will, of course, influence F_{CO_2}.

1.5.3.3. Open Systems

Open systems (Fig. 1.11) draw air of a known and controlled CO_2 and water vapour concentration from outside the system through a chamber enclosing the leaf or plant. In such a system the IRGA is calibrated in differential mode. A sample of the incoming air stream is passed through the reference tube and the air leaving the chamber is passed through the analysis tubes. Thus the IRGA measures the difference in the CO_2 content of the air before and after it passes through the leaf chamber.

$$F_{CO_2} = \frac{f \Delta C_a}{A} \qquad (1.5.3)$$

where f = the flow rate of air through the leaf chamber,

ΔC_a = the difference in CO_2 concentration before and after passing through the leaf chamber.

To determine F_{CO_2} in an open system the requirements are: that the IRGA can be used and is calibrated in differential mode, that the flow rate of air through the leaf chamber is constant and accurately known and that the leaf area is accurately determined. The main disadvantage of such a system is the initial expense, the requirement of an air-conditioning system, and the requirement of an IRGA which can accurately sense small differences in CO_2 concentration between two air streams, i.e. of the order of $1 \, \text{mg m}^{-3}$. There are a number of advantages of such a system. Firstly, by use of a switching device F_{CO_2} can be simultaneously determined for a number of chambers. Secondly, the CO_2, O_2 and water vapour concentration around the leaf can be easily manipulated. Thirdly, by linking an H_2O IRGA or differential psychrometer in series with the CO_2 IRGA, transpiration and photosynthetic CO_2 assimilation can be measured simultaneously.

1.5.3.4. Air Conditioning

An essential part of any system is a means of controlling the concentration of gases entering the chamber, particularly CO_2, O_2 and water vapour.

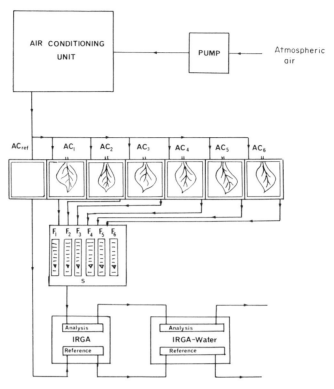

FIG. 1.11. Diagram of an open gas-exchange system. AC_n = Assimilation chambers; F_n = Flow meters; IRGA and IRGA-Water = Infra-red CO_2 and water vapour analysers, respectively; S = Gas sample selector. Arrows indicate direction of air flow.

(a) CO_2

Since CO_2 concentration is often limiting to the rate of CO_2 assimilation, its precise control is essential. In a closed system CO_2 cannot be controlled whilst in the semi-closed system CO_2 concentration is controlled by definition. However, in open systems control of CO_2 concentration becomes far more complex. The flow of air through an open system makes the continuous use of cylinders of compressed air containing known amounts of CO_2 impractical. Where only the atmospheric CO_2 concentration is required (604 mg m^{-3} of CO_2 in air at 20°C) air from outside the system could be used. However, the inlet must be distant from any source of CO_2, i.e. chimneys, combustion engines and living people! The roof of a 3+-storey building is often a good site for an inlet.

Even when distant from obvious sources of CO_2 pollution small fluctuations in atmospheric CO_2 concentration occur. These can be removed or damped by passing the inlet air through two or three containers of 0.5–1 m^3, e.g. well-cleaned oil drums. Where a range of CO_2 concentrations is required it is simplest to remove CO_2 from the inlet air by passing it through columns of KOH followed by a column of self-indicating soda lime (follow correct safety procedures when using KOH). Known amounts of CO_2 can then be added to the scrubbed air with a precision gas-mixing pump.

(b) O_2

O_2 cannot be controlled in a closed system, and in a semi-closed system care must be taken that the O_2 evolved does not increase the system concentration to the extent that photorespiration will be increased. Where a range of O_2 concentrations is required it is usually simplest to use prepared cylinders or, for low flow rates, a gas-mixing pump.

(c) H_2O

Water-vapour concentration has an important influence on stomatal opening. The water-vapour-pressure deficits (VPD) that can be withstood vary from species to species. In many mesophytes stomatal closure begins at a VPD of 1.0 kPa. Thus, care should be taken that VPD does not become inhibiting. Humidity within any system can be controlled by bubbling the air through water at a known temperature. A more efficient system is to pass the air first through water well above the required dewpoint and then bubble the air through water at the required temperature. This second bubbler acts as a condenser.

1.5.3.5. Assimilation Chamber

In general it is desirable that the environmental conditions, affecting photosynthesis, to which the leaf is exposed should be known to, and under the control of, the investigator. It is also essential that environmental variations across and along the leaf are kept to a minimum. Chamber design must take these two prerequisites into account with respect to the tissue under investigation, since a chamber designed for the leaves, or a leaf, of one species will not be suitable for many others. Chamber design must take account of the need to control ambient gas concentrations, leaf temperature and photon flux density.

(a) Ambient Gas Concentrations

The air-conditioning unit should supply known CO_2, O_2 and water-vapour concentrations to the chamber; however, since the leaf alters these concentrations care must be taken to prevent the build-up of significant concentration gradients. This can be achieved either by stirring the air with a fan within the chamber or by rapidly recirculating the air with a pump outside the chamber. Even when a ventilation technique is used, care should still be taken in design to avoid the creation of pockets of still air which can occur in corners or where a leaf is in close proximity to the chamber wall.

(b) Temperature

From a physiological standpoint it is the temperature of the leaf rather than the air that is of interest. Control of leaf temperature is greatly facilitated by minimizing the thermal radiation which reaches the leaf from the light source. The use of a long wave IR transmitting window in the chamber, such as polypropylene film (e.g. Propafilm C, I.C.I.), is useful in preventing a "green-house" effect within the chamber. Good chamber ventilation keeps leaf–air temperature gradients to a minimum and minimizes temperature gradients along and across the leaf. Leaf temperature is most commonly controlled by regulating the temperature of the ambient air in the chamber. This can be controlled either by jacketing the chamber so that coolant can be circulated over the chamber walls, by inserting cooling coils inside the chamber, or by building Peltier modules into the chamber walls (Jarvis et al., 1971).

(c) Photon Flux Density

A criterion in selection of light sources used for controlled environment CO_2 exchange studies is similarity to natural daylight. Xenon-arc lamps provide a close match, but this includes a high heat output. A combination of high-pressure Na and Hg lamps (1000 W) can also give a reasonable spectral match to daylight without the same heat output. It is still helpful to use a running water filter (10 cm depth) to remove thermal radiation. Photon flux density can be varied at the level of the chamber by interposing neutral density filters or even sheets of muslin. This is preferable to reducing the voltage supply to the lamps as this will alter the spectral composition, as well as quantity, of emitted radiation.

Only by using a perfectly spherical chamber with reflective walls and light entering through an inserted optical pipe could the radiation supplied to the leaf be totally diffuse. Chambers are commonly designed to receive direct radiation on the upper surface. If the base of the chamber is painted with optically black paint then the radiation conditions of the leaf can be precisely defined. Care should be taken to keep the chamber window perfectly transparent, scratches and smears produce surprising reductions in photon flux density (I_p) at points on the leaf surface. Finally, the leaf should be held in the horizontal plane in such a chamber if all parts of the surface are to receive the same I_p. This can be achieved by placing the leaf between two coarse meshes of fine transparent nylon.

1.5.4. Photosynthesis as a Diffusion Process

During photosynthesis CO_2 enters the leaf because a diffusion gradient exists between the sites of photosynthesis and the atmosphere. The net rate of photosynthetic CO_2 assimilation is the rate of CO_2 flux through this diffusion gradient. Flux (F_{CO_2}) is determined by the size of the gradient and the total resistance to CO_2 diffusion along that gradient. Flux of gases between regions of concentration difference is analogous to the flow of electricity through an electrical conductor. By analogy to Ohm's Law

$$F_{CO_2} = \frac{\Delta C}{\Sigma R} \qquad (1.5.4)$$

where the flux of CO_2 into the leaf (F_{CO_2}), the CO_2 concentration gradient (ΔC) and total resistance of the leaf to CO_2 diffusion (ΣR) are analogous to the current, potential difference and electrical resistance, respectively.

The idea of describing the process of CO_2 assimilation through a resistance analogue was first proposed by Browne and Escombe (1900). However, it was Gaastra (1959) who provided the first demonstrations of its use in practice. Gaastra (1959) considered that the pathway of CO_2 diffusion between the atmosphere and point of carboxylation consisted of three resistors in series: the boundary-layer resistance (r_a), the stomatal resistance (r_s) and the mesophyll resistance (r_m) (Fig. 1.12). Thus by expanding equa-

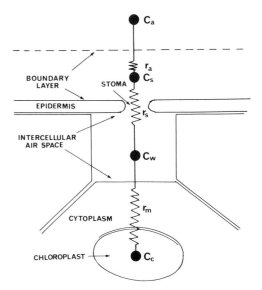

FIG. 1.12. Diagram illustrating a resistance analogue model of CO_2 diffusion into the leaf. $C = CO_2$ concentration (a = in ambient air, s = at the stomatal pore entrance, w = at the liquid/air interface of the mesophyll and c = at the site of carboxylation). r = resistance to CO_2 diffusion (a = in the boundary layer, s = through the stoma, m = in the mesophyll). F_{CO_2} is the flux of CO_2 through the chain of resistors, i.e. $F_{CO_2} = (C_a - C_s)/r_a = (C_s - C_w)/r_s = (C_w - C_c)/r_m = (C_a - C_c)/(r_a + r_s + r_m)$.

tion (1.5.4)

$$F_{CO_2} = \frac{C_a - \Gamma}{r_a + r_s + r_m} \qquad (1.5.5)$$

where C_a = the CO_2 concentration in the atmosphere.

The CO_2 concentration at the site of carboxylation is unknown, but is assumed to approach zero in Gaastra's model, thus $\Delta C \simeq C_a$. In later models the CO_2 compensation point of photosynthesis (Γ) has been considered a better estimate of the concentration inside the leaf.

1.5.4.1. The Nature of the Resistances

(a) *Boundary layer.* When a gas passes over a flat surface such as a leaf there is a small non-turbulent layer of air molecules associated with the surface. This is the boundary layer. The depth depends on the geometry of the surface and on the velocity of the gas flowing over the surface.

When the layer is deep, with a large surface or in still air, the resistance to gas diffusion is greater, the diffusion of water or CO_2 into and out of the leaf is slower and the resistance (r_a) is larger. r_a is decreased by increase in wind speed and decrease in leaf size. r_a is generally between 10 and 30 s m^{-1} and a small fraction of the total resistance.

Values can be determined using suitable mathematical formulae from rates of water loss, or of heat loss, from surfaces.

(b) *Stomatal.* The diffusive resistance encountered by CO_2 entering the leaf is due largely to the stomata. These represent the variable resistance in the non-metabolic part of the photosynthetic CO_2 assimilation pathway. Calculation of the stomatal resistance is thus of great importance and is dealt with in Section 1.7.

(c) *Mesophyll.* This is really a combination of resistances which includes physical diffusion resistances to movement of CO_2 in the air spaces and from the mesophyll cell walls to the sites of carboxylation together with biochemical rate limitations. In practise r_m can be determined by subtracting r_a and r_s estimated from water vapour transfer, from the total resistance (Σr).

Where:

$\Sigma r = (C_a - \Gamma)/F_{CO_2}$,
$\Gamma = CO_2$ compensation point of photosynthesis,
$r_m = \Sigma r - (r_a + r_s)$.

1.5.4.2. Estimation of Diffusive Resistances

(a) *Boundary-layer resistance.* A number of techniques have been described for determining this resistance (Jarvis, 1971). The simplest of these is the measurement of the rate of water-vapour loss from a moist filter-paper replica of the leaf. Providing that the surface of the paper is wet the only resistance to water-vapour loss is the boundary-layer resistance.

$$r_a^{H_2O} = \frac{\chi_s - \chi_a}{F_{H_2O}}$$

where $r_a H_2O$ = resistance to water vapour transfer across the boundary layer,

χ_s = water vapour concentration determined at the saturated content of air at the leaf temperature (Goff, 1965),
χ_a = the ambient water vapour concentration,
F_{H_2O} = the rate of water-vapour loss or evaporation.

Since CO_2 molecules are larger than water molecules, the resistance for CO_2 will be larger than that for water vapour. If transfer was purely by diffusion the ratio of the two resistances would be directly proportional to the ratio of the diffusion coefficients. However, transfer through the boundary layer is part by diffusion and part by turbulent mixing, the following conversion is derived empirically to take account of this problem (Thom, 1968).

$$r_a = r_a^{H_2O} \times 1.387.$$

(a) *Stomatal resistance* (r_s). Determination of r_s (and its reciprocal stomatal conductance, g_s) has been considered in detail in Section 1.7.

In gas-analysis systems the most common procedure is to determine the stomatal resistance to water-vapour transfer from the leaf and then estimate the resistance to CO_2 transfer, from the ratio of the molecular diffusion coefficients of CO_2 to water vapour, i.e. 1.605 (Jarvis, 1971).

$$r_s^{H_2O} = \frac{\chi_s - \chi_a}{F_{H_2O}} - r_a^{H_2O},$$

$$r_s = r_s^{H_2O} \times 1.605$$

where $D_{H_2O}/D_{CO_2} = 1.605$
and D_{CO_2} = diffusion coefficient of CO_2 molecules,
D_{H_2O} = diffusion coefficient of water vapour molecules,
$r_g = r_s + r_a$.

(c) *Mesophyll resistance* (r_m). This resistance can be determined by difference. Since by rearranging equation (1.5.5)

$$r_m = (C_a - \Gamma)/F_{CO_2} - (r_a + r_s). \qquad (1.5.6)$$

1.5.4.3. Application of Resistance Models

Resistance models allow numerical evaluation of limitation to CO_2 diffusion and thus productivity, given that supply of CO_2 is limiting photosynthesis. Thus, if $r_s = 250$ s m^{-1} and $\Sigma r = 500$ s m^{-1} it can be said that r_s accounts for 250/500 or one-half of the sum of limitations to CO_2 assimilation in that leaf under the conditions of measurement.

Typical resistances to carbon dioxide diffusion for a leaf of a mesophyte under optimal conditions would be in the ranges:

$r_a = 10$–30 s m^{-1},
$r_s = 250$–1000 s m^{-1},
$r_m = 250$–4000 s m^{-1} (25–500 s m^{-1}, C_4).

Thus even when fully open the stomata normally represent the largest resistance in the gaseous diffusion vapour pathway to water and therefore for many purposes $r_s \simeq r_g$. Indeed, r_s often represents the major limitation to CO_2 assimilation for leaves of C_4 plants in full sunlight, where r_s may typically represent 80–90% of the total resistance to CO_2 assimilation. Thus, in many tropical environments stomata may present the largest single limitation to conversion of solar energy into chemical energy stored in carbohydrates by higher plants.

Bibliography and Further Reading

BANWELL, C. N. (1966) *Fundamentals of Molecular Spectroscopy.* (McGraw-Hill: London.)

BISCOE, P. V., CLARK, J. A., GREGSON, K., MCGOWAN, M., MONTEITH, J. L. and SCOTT, R. K. (1975) Barley and its environment: 1. Theory and practice. *J. appl. Ecol.* **12**, 227–247.

BROWN, H. T. and ESCOMBE, F. (1900) Static diffusion of gases and liquids in relation to the assimilation and translocation in plants. *Phil. Trans. Roy. Soc.* [B]**193**, 223–291.

GAASTRA, P. (1959) Photosynthesis of crop plants as influenced by light, carbon dioxide, temperature and stomatal diffusion resistance. *Meded. Landb. Hogesch. Wageningen* **59**, 1–68.

GOFF, J. A. (1965) Saturation pressure of water on the new Kelvin Scale. In: *Humidity and Moisture,* Vol. 3 (Ed. by A. WEXLER), pp. 289–292. (Reinhold Corp.: New York.)

HILL, D. W. and POWELL, T. (1968) *Non-dispersive Infra-red Gas Analysis in Science, Medicine and Industry.* (Adam Hilger Ltd.: London.) 222 pp.

JANÁČ, J., ČATSKÝ, J. and JARVIS, P. G. (1971) Infra-red gas analysers and other physical analysers. In: *Plant Photosynthetic Production. Manual of Methods* (Ed. by Z. ŠESTÁK, J. ČATSKÝ and P. G. JARVIS), pp. 111–193. (Dr. W. Junk: The Hague.)

JARVIS, P. G. (1971) The estimation of resistances to carbon dioxide transfer. In: *Plant Photosynthetic Production. Manual of Methods* (Ed. by Z. ŠESTÁK, J. ČATSKÝ and P. G. JARVIS), pp. 566–622. (Dr. W. Junk N.V.: The Hague.)

JARVIS, P. G., ČATSKÝ, J., ECKARDT, F. E., KOCH, W. and KOLLER, D. (1971) General principles of gasometric methods and the main aspects of installation design. In: *Plant Photosynthetic Production. Manual of Methods* (Ed. by Z. ŠESTÁK, J. ČATSKÝ and P. G. JARVIS), pp. 49–110. (Dr. W. Junk: The Hague.)

MARTIN, T. J., SWAN, A. G. and RAWSON, H. M. (1974) An air conditioner for leaf chamber gas exchange studies providing accurate control of temperature and relative humidity and a recorder output of the latter. *Photosynthetica* **8**, 216–220.

MONTEITH, J. L. (ed.) (1975) *Vegetation and the Atmosphere.* (Academic Press: London.) 2 vols.

PARKINSON, K. J. and LEGG, B. J. (1971) A new method for calibrating infra-red gas analysers. *J. Phys. E. Sci. Instrum.* **4**, 598–600.

THOM, A. S. (1968) The exchange of momentum, mass, and heat between an artificial leaf and the airflow in a wind-tunnel. *Quart. J. Roy. Met. Soc.* **94**, 44–55.

Appendix 1.5. Experimental Work: Photosynthetic Gas Exchange

Experiment

An experiment which is used in the training courses and can be completed by a group of students within a half-day is the determination of light utilization efficiency in photosynthesis by an attached leaf in a range of photon flux densities.

Objectives

The objectives are to illustrate the following points to the students: The operation, construction, and calibration of an infra-red gas analyser, the construction of simple closed and open gas-exchange systems, absolute and differential measurement of CO_2 concentrations with an IRGA, calculation of the rate of CO_2

assimilation (F_c) and the apparent quantum efficiency. Finally, sources of error in gas analysis studies.

Equipment Required

A CO_2 IRGA suitable for operation in both absolute and differential mode, connected to a pen recorder.

At least two temperature-controlled leaf chambers.

A cooled artificial light source able to provide a photon flux density of $1800\ \mu mol\ m^{-2}\ s^{-1}$ at the chamber surface.

Neutral density filters, such as sheets of white muslin or white fibre-glass of various thicknesses.

A simple open-circuit gas-exchange system as described on page 29, except that the sample selector is not essential.

A cylinder of calibration gas and, if available, a gas diluter (e.g. GD600, Analytical Development Co.) or a gas mixing pump (e.g. SA27/2f, Wösthoff GmbH).

A supply of CO_2-free air (note that some IRGAs, e.g. Series 225/3 Analytical Development Co., include a built-in supply) or of pure N_2.

Thermocouples, suitable for leaf temperature measurement.

A quantum sensor and meter.

Graph paper and calculators.

Plant Material

In the training courses leaves ranging from the fronds of bearing coconut trees to the leaves of summer annual grasses such as *Digitaria sanguinalis* have been used. Mature leaves of 3-week-old *Phaseolus vulgaris* and 5-week-old *Amaranthus edulis* have proved the most successful, although almost any healthy leaves are suitable except succulents and other species which could utilise C.A.M. and thus may show nocturnal stomatal opening. *Phaseolus vulgaris* and *A. edulis* can be used in the same leaf chambers and provide a comparison between a C_3 and a C_4 crop, respectively. These plants are grown without check, i.e. in a well- and regularly watered potting compost such as John Innes No. 2, at about 25°C and in good light.

Procedure

The group (ideally of no more than five students if they are to obtain "hands-on" experience) will be shown the construction of the IRGA and the principle of its operation is explained. The IRGA will then be calibrated in absolute mode using the calibration cylinder and, if available, a gas diluter or a gas mixing pump. The calibration will be demonstrated once. Now proceed as follows:

Deliberately upset the zero and gain settings, and as a group re-calibrate the IRGA (help will be given if necessary). Now pump laboratory air through the analysis cell. This will demonstrate the high and fluctuating CO_2 concentrations encountered in laboratories and emphasize the need to collect air for open systems at some distance from the laboratory. You (the students) should now measure the area of the leaf to be placed in the leaf chamber. The leaf outline should be traced onto graph paper, cut out, weighed, and the area determined from the weight/area ratio for the graph paper.

Now seal the leaf into the chamber and make a simple closed system (page 28). You should check for leaks in the system by breathing onto individual tubing joints and seals. Exhaled breath contains about $100,000\ mg\ m^{-3}$ of CO_2 and thus provides a simple and effective gas for detecting leaks in a system capable of resolving CO_2 concentration changes of less than $2\ mg\ m^{-3}$. It is instructive to deliberately leave one or two joints loose.

Once the system has been made gas-tight, record the rate of CO_2 depletion in the system. Note the time taken for the concentration to decrease from about 600 to $580\ mg\ m^{-3}$. Calculate F_c (Eq. 1.5.1, page 28) whilst waiting for the CO_2 concentration to fall to the compensation point (an estimate of the system volume will be provided).

Next reconnect the chamber so that it is in open system, with a second empty chamber as a reference. Firstly, measure the CO_2 concentration reduction caused by the leaf, using the IRGA in absolute mode. Compare the CO_2 concentrations in the leaf and reference gas streams by passing each in turn through the analysis cell of the IRGA;

this difference at best will probably be no more than 1–2% of F.S.D. with most IRGAs. A differential calibration will now be conducted. Connect the leaf and reference chambers to the analysis and reference cells of the IRGA, respectively. Use in differential mode will amplify the difference between the two gas streams and provide a good demonstration of the value of differential gas analysis. Recalculate F_c using Eq. 1.5.3 (page 29). You will probably find that the F_c calculated in open system is higher than in closed system for the same leaf, since a steady-state rate could not be obtained in the latter.

Reduce the photon flux density (I_p) in the leaf chamber by interposing neutral density filters between the light source and the leaf. Once the leaf has reached a new steady-state F_c, I_p may be further reduced. Filters should be used to provide ten to twenty values of I_p from about 2000 μ mol m^{-2} s^{-1} down to 0. Since the most rapid changes in F_c occur at the lowest values of I_p, make sure that low light levels are well represented. As I_p is decreased, the leaf will cool and this must be compensated for by increasing the chamber air temperature, so that the leaf temperature is held constant.

Graph the response of F_c against I_p. If F_c is recalculated in μ mol m^{-2} s^{-1}, then the efficiency of light conversion or apparent quantum efficiency may be calculated ($\Phi = F_c/I_p$) and plotted against I_p. This will show how efficiency decreases with increase in I_p. Plots of Φ against I_p for P. vulgaris (C$_3$) and A. edulis (C$_4$) provide a good comparison in the training courses. Phaseolus vulgaris shows higher efficiencies at low light levels whilst A. edulis shows almost double the efficiency of P. vulgaris at light levels close to that of full sunlight.

A number of simple extensions of this work are possible. Spectral filters may be employed to show the effects of light quality on quantum efficiency and light energy conversion efficiency. If the CO$_2$ is removed from the air entering the system by passing it through columns of soda lime a rate of photorespiration can be measured. The effect of a high water-vapour pressure deficit may be illustrated by drying the air with a column of dry CaCl$_2$ granules before it enters the leaf chamber.

1.6. FIELD PHOTOSYNTHESIS; MONITORING WITH ^{14}CO$_2$

by J.-E. HÄLLGREN

1.6.1. Introduction

In recent years there has been an increasing interest in the measurement of photosynthetic rates of plants in the field. However, many scientists cannot afford the expensive equipment generally used and, moreover, you may work at experimental sites which do not have power supplies, or are inaccessible to vehicles for use in the field. Here the availability of a simple and cheap ^{14}CO$_2$ (see Appendix 1.6.2 for definition of symbols used) technique may be of interest.

The subject of measuring photosynthesis using ^{14}CO$_2$ techniques has been reviewed by Voznesenskii et al. (1971) and more recently by Incoll (1977).

The methods actually fall into two somewhat different groups. In the first, the leaf or other photosynthetic tissue is exposed to a gas mixture in which the carbon dioxide concentration consists of a mixture of ^{14}CO$_2$ and ^{12}CO$_2$. After a certain time of exposure, the leaf/tissue is killed and the amount of ^{14}C it has fixed is determined.

The rate of photosynthesis may then be estimated if it is assumed that the two isotopes of carbon have been photosynthetically fixed in proportion to their partial pressures in the air surrounding the leaf/tissue during the incubation.

In the second group of methods, the leaf is exposed in a closed space to a gas mixture containing the carbon isotopes, and the rate of decrease of radioactivity of the gas is measured using a β-counter built into the closed space.

Apart from errors such as technical limitations and instrumental difficulties both these groups of methods suffer from some types of errors, i.e.

1. isotope discrimination,
2. dilution by respiratory ^{12}CO$_2$,
3. higher diffusion rate ^{14}CO$_2$ at the start,
4. air spaces in leaves may trap ^{14}CO$_2$ (unassimilated).

Several successful field systems have now been developed (Austin and Longden, 1967; Strebeyko, 1967; Shimshi, 1969; Incoll and Wright, 1969 and Tieszen et al., 1974).

Basically, any $^{14}CO_2$ system must successfully employ five stages.

1. A delivery system which provides constant flow rates of air containing known absolute concentrations of CO_2 with known specific activities of $^{14}CO_2$.
2. An assimilation chamber which ensures a minimum of disturbance to the plant, and as little alteration of irradiance and temperature as possible.
3. Prevention of the loss of ^{14}C from plant material following exposure to $^{14}CO_2$.
4. A convenient sample combustion process for radioactivity determination, e.g. scintillation counting.
5. A radioactivity counting process that provides for a reliable and highly efficient measurement of ^{14}C content.

The simplest field system is a gas cylinder containing the gas mixture of $^{14}CO_2$ and $^{12}CO_2$, connected to a small exposure chamber by a gas tubing line (Fig. 1.13). The flow rate is controlled (not critical as long as it does not limit photosynthesis). The plant material is exposed for an accurately measured time. The sample is cut, the tissue killed (in liquid nitrogen) and prepared for counting. When the radioactivity has been assayed (e.g. by liquid scintillation counting) the photosynthetic rate can be calculated.

$$F_{CO_2} = \frac{ASSAY}{A \cdot t \cdot E \cdot SR}$$

t = time of exposure,
E = efficiency of counting,
SR = specific activity of the gas passing over the leaf,
$ASSAY$ = measured radioactivity (Bq − counts s^{-1}).

1.6.2. A Two-Gas System

There are some technical problems with the system described above, which have led to the development of a two-gas system described in detail by Incoll (cf. Fig. 1.14).

This system overcomes some major problems with the one gas system, i.e. dilution of the specific radioactivity, timing errors and wasting of labelled gas. This system incorporates an additional unlabelled gas stream and a gas-samp-

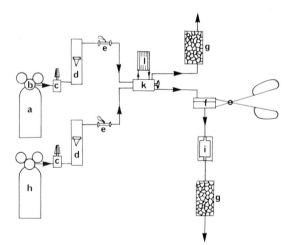

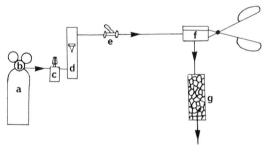

FIG. 1.13. Schematic diagram of a simple field apparatus for exposing plant tissue to $^{14}CO_2$: (a) gas cylinder containing a mixture of $^{14}CO_2$ and $^{12}CO_2$ in air; (b) single-stage pressure regulator; (c) flow regulator; (d) variable-area flow meter; (e) on–off toggle valve; (f) leaf exposure chamber; (g) absorbent column. (a) to (e) are joined by narrow-bore stainless-steel tubing, (e) to (g) by flexible PVC or butyl rubber tubing (after Incoll, 1977).

FIG. 1.14. Schematic diagram of a two-gas field system for exposing plant tissue to $^{14}CO_2$. (a–g) see legend of Fig. 1.13; (h) gas cylinder containing $^{12}CO_2$ in air; (i) water-vapour sensor, temperature controlled; (k) gas-sampling valve with (l) stainless-steel loop (10 cm^3 capacity). All components except (f), (i) and (g) are mounted on one back-carrying-frame. The hygrometer electronics are mounted on another carrying-frame with an electronic thermometer and a quantum sensor meter (after Incoll, 1977).

ling valve. The valve dispenses an aliquot (10 cm^3) of the radioactive gas into the non-radioactive gas stream.

In operation, the exposure chamber is clamped on the leaf and unlabelled gas is passed over the leaf. Simultaneously labelled gas from the other field cylinder is passed through the sample loop. The valve plunger is then pushed in and a pulse of labelled gas is carried to the leaf chamber. The pulse takes a finite time to pass the chamber which is a function of flow rate and volume of tubing (Incoll, 1977).

To this system a steady state humidity sensor can be added into the effluent gas stream. Such a sensor can be used as a porometer.

In the two-gas system the photosynthetic rate is calculated as

$$F_{CO_2} = \frac{ASSAY \cdot f}{A \cdot E \cdot SR \cdot V_d}.$$

Hence, you need to know f and V_d accurately.

f—the flow rate is controlled precisely with a flow regulator, and monitored with a flowmeter.

V_d—volume of loop—the valve dispenses an accurate volume of, for example, 10 cm^3.

1.6.3. ^{14}C Porometer and Double Isotope Porometer

It is obviously a great advantage to be able to measure the photosynthetic rate and the stomatal resistence concurrently on the same leaf. Bravdo (1972) was the first to describe an ingenious modification of a ventilated diffusion porometer (cf. M. M. Ludlow, Measurement of stomatal conductance and plant water status, Section 1.7) which enabled simultaneous measurements of leaf resistance and photosynthesis using ^{14}CO$_2$. Concurrent measurements can also be made in the apparatus described above by using the humidity sensor and making use of the period when unlabelled gas is flowing.

An important new alternative for making simultaneous determinations of transpiration and photosynthesis under field (and laboratory) conditions has recently been published (Johnson et al., 1979).

The porometer recently developed employs two radioactive isotopes, tritium and ^{14}C in a gaseous phase. ^{14}C is administered as ^{14}CO$_2$ in a

manner basically similar to that described by Shimshi (1969). Tritium (T) is administered simultaneously as ^{3}H$_2$O vapour in the same air stream. This is achieved by bubbling ^{14}CO$_2$ labelled air from a pressurized cylinder through a ^{3}H$_2$O supply held at a constant temperature in an ice bath. At the assimilation chamber clamp, a finger release spring valve is used to start and stop the flow of isotopes through the leaf chamber. Capillary tubing between the valve and the exposure chamber controls the flow rate of the isotope gas mixture. The leaf or plant part enclosed in the assimilation chamber is hence exposed to ^{14}CO$_2$ and ^{3}H$_2$O vapour when the gas flow is initiated with the finger release valve. Sampling time is in the order of one measurement per minute. A correct understanding of the actual concentrations of T and ^{14}C in the air stream entering the assimilation chamber is crucial to subsequent evaluations. These values can be checked by withdrawing the air stream at the chamber inlet directly into a syringe. Syringe samples are then bubbled into evacuated self-sealing stoppered scintillation vials containing trap materials, such as phenethylamine for ^{14}C and H$_2$O for ^{3}H$_2$O. The total conductance to water vapour and CO$_2$ can be calculated directly according to the formula

$g_{cm\,s^{-1}} =$

$$\frac{\text{(radioactive disintegrations cm}^{-2}\,\text{s}^{-1}\text{) uptake}}{\text{(radioactive disintegrations cm}^{-3}\text{) air stream}}.$$

The double isotope porometer is portable and can therefore be used for extensive field studies. Note that ^{3}H$_2$O may exchange with leaf surface-bound water and thus must be accounted for.

1.6.4. Preparation of Gas Mixtures

There are several alternative methods available, of which one will be described here. What you actually need is seen in Fig. 1.15(a). Compressed air from a cylinder is led through a steel cylinder containing an absorbent for CO$_2$(2). The CO$_2$ free air from the cylinder blows an acid (1 M H$_2$SO$_4$ 10 cm^3) down to a big glass test tube, containing a known mixture of NaH^{14}CO$_3$/NaH^{12}CO$_3$. Hence a known amount of CO$_2$ will be evolved, and the air containing

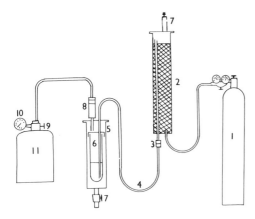

FIG. 1.15(a). A simple system for preparation of a gas mixture of $^{14}CO_2$ and $^{12}CO_2$. 1. high-pressure gas cylinder containing air with regulatory valve; 2. steel cylinder with CO_2 absorbent (carbosorb or NaOH); 3. gas-tight connection; 4. tubing in which H_2SO_4 is added; 5. steel cylinder with removable lid; 6. glass test tube containing $NaH^{14}CO_3/NaH^{12}CO_3$ mixture; 7. safety valves; 8. back valve; 9. regulatory valve; 10. pressure gauge; 11. aluminium field cylinder to be filled.

FIG. 1.15(b). A simple wet oxidizing method. Two scintillation vials, connected with plastic tubing, used for wet combustion of leaf disks and absorption of the evolved CO_2 in NaOH. (a) Vial with leaf disk and added chromic acid (10–20 ml). (b) Vial with 0.25 M NaOH. NaOH. The scintillation cocktail (e.g. In-stagel) is added to vial b after 24 hours and the vial is counted in a scintillation counter.

$^{14}CO_2/^{12}CO_2$ is led into an aluminium field cylinder. The cylinder is filled until a predetermined pressure is achieved.

An example of the calculation is given below.

1. You wish to prepare a gas mixture e.g. 350 ppm CO_2 in air.
2. The specific activity is 0.5 Ci mol^{-1}.
3. The volume of the aluminium field cylinder is 14,300 cm^3 (water capacity) and should not be filled to a pressure higher than 10–12 bar.

4. $P \cdot V = n \cdot R \cdot T$ gives you how many moles CO_2 each bar increase in pressure corresponds to. Therefore when

$P = 1\,bar$
$V = 14,300\ cm^3,$
$T = 295\ K,$
$R = 83.144\ cm^3\ bar\ K^{-1}\ mol^{-1},$
$n_{CO_2} = 350 \times 10^{-6} \times n_{air},$
$n_{CO_2} = 350 \times 10^{-6} \times \dfrac{1 \times 14,300}{83.144 \times 295}$
$\quad = 204.06 \times 10^{-6}\ mol.$

This means that if you fill the field flask to 10 bar, the gas bottle will contain 2.0406 mmol CO_2.

Radioactive carbonate solutions can be bought in ampoules containing, for example, 1.0 mCi, with a specific activity of 0.5 mCi mmol^{-1}. Thus this ampoule corresponds to 2 mmol CO_2. This will, as seen above, correspond to a total pressure in the gas cylinder of

$$\frac{2}{0.20406} = 9.80\ bar.$$

How much non-radioactive carbonate has then to be added?

EXAMPLE: 1 ampoule contains:

radioactivity 1 mCi,
amount 1.4 mg = 0.016 66 mmol,
volume 1.0 ml.

Hence you have to add

$2 - 0.0167 = 1983$ mmol $NaH^{12}CO_3$.

Technology
1. Prepare a 1.0 M solution of $NaH^{12}CO_3$ (840 mg $NaHCO_3$ is dissolved in 10.0 ml of H_2O with the pH adjusted to 8 with NaOH).
2. Add to the big glass test tube
   (a) $NaHCO_3$-solution, i.e. 1.98 ml in the example above,
   (b) 1 ampoule of $NaH^{14}CO_3$ (1.0 mCi, 1.0 ml),
   (c) a small amount of olive oil.
3. Fill the field cylinder to the calculated pressure (9.80 bar). **Observe** that the meter gives you the overpressure. If the air in the field flask is totally free from CO_2, you have to fill it up to 8.80 bar.

4. Check the gas mixture by taking out a known volume and dissolve it in some absorbent, and analyse it. If possible analyse it in an IR-gas analyser calibrated in the absolute mode (page 27).

General Aspects About the Gas Mixture

The concentration of $^{14}CO_2$ and $^{12}CO_2$ must be accurately known so that reliable values of F_{CO_2} are obtained. In some countries it is possible to buy the gases from commercial supplies with reliable analyses of $^{14}CO_2$ and $^{12}CO_2$ concentrations. It must be appreciated that commonly used steel cylinders may adsorb CO_2 to a small extent whereas aluminium ones do not. Therefore it is recommended that you use aluminium field cylinders.

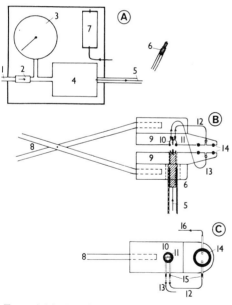

FIG. 1.16. Details of the apparatus for measuring photosynthesis. (A) Manometer and air reservoir unit: (1) connection to compressed air cylinder, (2) one-way valve, (3) manometer, (4) air reservoir, (5) tube to leaf clamp, (6) tyre valve, (7) CO_2-absorber. (B and C) Leaf clamp: (8) pair of tongs, (9) perspex plates, (10) adjustable screw, (11) rubber gasket ring, (12) plastic tube, (13) plastic tube, (14) rubber gasket ring, (15) capillary, (16) connection to CO_2-absorber.

1.6.5. The Assimilation Chamber

There are several descriptions of assimilation chambers in the literature; however, it is recommended that you build a chamber that suits your own plant material to be examined (see page 31). The assimilation chamber has to be constructed so that it minimizes mechanical disturbance of the leaf. A construction of plexiglass is generally easy to build and work with. A suitable design for a small chamber is shown in Fig. 1.16.

1.6.6. Methods for Killing and Storing Your Material

The leaf is killed by placing it into liquid nitrogen (solid carbon dioxide might do as well) and then put it in a premarked envelope which is stored in a freezing bag together with ice lumps. Another possibility is to punch the leafpiece that has been incorporated with $^{14}CO_2$ directly into a scintillation vial, which contains a tissue solvent. The general principle is: the simpler the method is in the field the more samples can be taken in a given time.

1.6.7. Oxidizing the Plant Material

The simplest way (and the most expensive one) to oxidize your plant material is to use an automatic sample oxidizer (Packard or ICN).

A simpler but more time-consuming method is to oxidize the leaf sample with dichromate–sulphuric acid, and trap the evolved CO_2 in sodium hydroxide. A simple set up is shown in Fig. 1.15(b). (Dichromate sulphuric acid: 63 g $Na_2Cr_2O_7 + 35$ cm^3 H_2O in 1000 cm^3 conc. H_2SO_4.)

An alternative procedure, if a wet digestion technique is used, would be to immediately place the plant material into a basic tissue solvent, e.g. Soluene (Packard Instruments).

Bibliography and Further Reading

AUSTIN, R. B. and LONGDEN, P. C. (1967) A rapid method for the measurement of rates of photosynthesis using $^{14}CO_2$. *Ann. Bot.* **31**, 245–253.
BRAVDO, B.-A. (1972) Photosynthesis, transpiration,

leaf stomatal and mesophyll resistance measurements by the use of a ventilated diffusion porometer. *Physiol. Plant.* **27**, 209–215.

INCOLL, L. D. (1977) Field studies of photosynthesis: Monitoring with $^{14}CO_2$. In: *Environmental Effects on Crop Physiology* (Ed. by LANDSBERG, J. J. and CUTTING, C. V.). (Academic Press: London, New York, San Francisco.)

INCOLL, L. D. and WRIGHT, W. H. (1969) A field technique for measuring photosynthesis using 14-carbon dioxide. *Conn. Agric. Exp. Sta Spec. Bull. Soils* 30/**100**, 1–10.

JOHNSON, H. B., ROWLANDS, P. G. and TING, I. P. (1979) Tritium and carbon-14 double isotope porometer for simultaneous measurements of transpiration and photosynthesis. *Photosynthetica* **13**(4), 409–418.

SHIMSHI, D. (1969) A rapid field method for measuring photosynthesis with labelled carbon dioxide. *J. Exp. Bot.* **20**, 381–401.

STREBEYKO, P. (1967) Rapid method for measuring photosynthetic rate using $^{14}CO_2$. *Photosynthetica* **1**, 45–49.

TIESZEN, L. L., JOHNSON, D. A. and CALDWELL, M. M. (1974) A portable system for the measurement of photosynthesis using 14-carbon dioxide. *Photosynthetica* **8**(3), 151–160.

VOZNESENSKII, V. L., ZALENSKII, O. V. and AUSTIN, R. B. (1971) Methods of measuring rates of photosynthesis using carbon-14 dioxide. In: *Plant Photosynthetic Production, Manual of Methods* (Ed. by ŠESTÁK, Z., ČATSKÝ, J. and JARVIS, P. G.). (Dr. W. Junk, N.V. Publishers: The Hague.)

Appendix 1.6.1. Experimental Work on $^{14}CO_2$ Fixation

Objective

Your objective is to measure the $^{14}CO_2$ incorporation of leaves at various heights in a C_4 (e.g. *Zea mays*) canopy, and a C_3 (e.g. *Pisum*) crop, and the corresponding quantum flux densities at these leaves. You will work in the field together with a second group which will provide data on stomatal conductance and other environmental factors (Appendix 1.7). You will therefore be in the position to correlate your photosynthetic measurements with leaf temperature and stomatal conductances of the leaves.

Procedure

Students will work in pairs, and each person will perform the incorporation of $^{14}CO_2$ into leaves, sample combustion, radioactivity counting and measurements of quantum flux densities.

The $^{14}CO_2$ Incorporation

The construction and principle of the apparatus and the preparation of radioactive gas will be described to all students in detail in advance. You will thereafter be able to use the equipment for your measurements, and each student will practice on the equipment themselves.

Sample Combustion

A simple direct in-vial sample oxidation will be used. An organic tissue solvent (e.g. Soluene) will be used, and should be handled with care. The radioactive part of the leaf is punched out and put into the vial with a pair of forceps.

Radioactivity Counting

A short technical description of the scintillation counter and the scintillation procedure will be given. The students will then count their samples and calculate the photosynthetic rate in terms of $\mu g\, CO_2\, m^{-2}\, s^{-1}$.

The Photosynthetic Active Radiation (PAR)

The PAR (400–700 nm) will be measured by a quantum sensor, at the position of the leaf, during the $^{14}CO_2$ incorporation. All data shall be recorded thoroughly. Photosynthetic rates, measured by $^{14}CO_2$ incorporation, are plotted versus the measured quantum flux densities incident on the leaves. Hence, each student must get noted and calculated data from the other students in the group, and plot the graphs in their own experimental notebooks.

Calculations

You can calculate the appropriate number of decompositions in the sample after quench corrections have been made. The calibration curve for the scintillation counting will be made up in advance by using either external standard, internal standard or the channels-ratio method described in the Manual.

Modern scintillation counters normally calculate the numbers of disintegrations per minute (dpm). Assume that, after appropriate quench corrections, you record a dpm in your sample.

A dpm = $A/60$ disintegrations per second (dps) = $A/60$ Bq.

The specific activity in the gas mixture must be known. Assume that the specific activity is B Ci mol^{-1}.

1 Ci corresponds to 222×10^{10} Bq

$$= \frac{A}{60 \times B \times 3.7 \times 10^{10}} \text{ mol CO}_2.$$

The molecular weight of CO_2 is 44 g.

$$\frac{A \times 44}{60 \times B \times 3.7 \times 10^{10}} \text{ g CO}_2.$$

The exposure time in seconds = S.
The exposed leaf area in cm^2 = $L \times 10^{-4}$ m^2.

$$\frac{A \times 44 \times 10^{-6}}{60 \times B \times 3.7 \times 10^{10} \times S \times L \times 10^{-4}} \mu\text{g m}^{-2}\text{ s}^{-1}.$$

Analysis of Data

Each student should compare the plots of incident quantum flux densities versus photosynthesis for the C_4 and C_3 plants respectively.

Discuss following questions:

What other parameters are of importance to fully interpret this type of plot?

Discuss the methodological difficulties and possible disadvantages/advantages with the $^{14}CO_2$ method.

Appendix 1.6.2. Symbols and Abbreviations Used

$^{14}CO_2$ radioactive isotope of carbon
$^{12}CO_2$ most abundant natural non-radioactive isotope of carbon
F_{CO_2} flux of carbon to the leaf
A leaf area (m^2)
E symbol used to express the efficiency of determining a radioactive isotope ($^{14}CO_2$) with the scintillation counter (see Scintillation counting)

ASSAY the measured radioactivity (the SI unit to be used is Bq)
Bq   Bequerel; 1 Bq = 1 disintegration per second = 1 dps
SR   specific radioactivity of the gas passing over the leaf. The specific radioactivity expresses the amount of radioactive isotopes per total amount of non-radioactive carbon isotopes (units Bq mol^{-1})
f flow rate of the gas (m^3 s^{-1})
V_d volume of loop (m^3)
T temperature in degrees Kelvin, K
R gas constant (units can be joules mole^{-1} K^{-1} or litre-atmosphere mole^{-1} K^{-1} or m^3 Pa mol^{-1} K^{-1} or cm^3 bar mole^{-1} K^{-1})
n_{CO_2} number of molecules of the gas (in this case CO_2)
mCi   millicurie; an older but still frequently used unit of expressing radioactivity based on Ci
Ci   Curie; 1 Ci = 3.7×10^{10} Bq, 1 mCi = 3.7×10^{7} Bq

Appendix 1.6.3. Liquid Scintillation Counting

The Scintillation Counter

A scintillator is a substance which emits a weak light flash when struck by an ionizing particle. The flashes are detected by a photomultiplier (in practice two set up in a coincidence mode to reduce noise) which amplifies the signal producing an electric pulse which is fed into electronic circuits for amplification, pulse-amplitude analysis, data processing and recording.

The counting rate of the electronic pulses is the same as that of the original scintillations detected, and this in turn is determined by the number of disintegrations per minute or activity of the source of ionizing radiations. The pulse amplitude is proportional to the scintillation intensity, and with beta-particle irradiation, this is proportional to the energy of the beta-particle which produced it. Hence the scintillation counter functions as a beta-spectrometer. This is an important property which is not possessed by other ionization detectors, such as Geiger-Muller counters, which do not distinguish between ionizing particles of different energies. The scintillation counter also has the advantage of a high efficiency, around 80 to 90% for ^{14}C.

The common isotopes used for biological work differ in their characteristic maximum β-particle energy (^3H, 18.6 keV; ^{14}C, 156.7 keV), hence it is possible to count two isotopes simultaneously.

The Scintillation Process

The liquid scintillation process occurs in a mixture termed the scintillation cocktail. This consists of several components, an aromatic solvent, a fluorescent aromatic solute, and the sample. The passage of an ionizing particle through the solution mainly causes ionization and excitation of the molecule of the solvent, since these form the bulk of the scintillator. As a result, a number of excited solvent molecules are produced. The energy is passed from molecule to molecule until transferred to a scintillator molecule resulting in fluorescence, which is detected as the scintillation emission.

Scintillators (Solute)

One of the most popular primary solutes is 2–5 diphenyl oxazole (PPO), which combines a high solubility with a high fluorescence quantum efficiency. In some cases a secondary solute such as 1,4,di(2-(5-phenyloxazolyl))-benzene (POPOP) is also included to aid matching between the fluorescence wavelength and the photomultiplier sensitivity.

Cocktails

An efficient liquid scintillator (cocktail) is toluene containing 4 g/l PPO, to which 0.1 g/l POPOP may be added. However, this may not be suitable for many biological materials, as only a small proportion of biological molecules (such as steroids, lipids, fatty acids, hydrocarbons and gases such as acetylene) will dissolve in it. Since water is not miscible with toluene a variety of techniques have been developed in order to enable water (and water-soluble compounds) to be counted. These include:

1. Chemical treatment (e.g. hyamine hydroxide) to render compound soluble in toluene.
2. Addition of a second solvent, e.g. dioxan (usually used with 50 to 100 g naphthalene per l).
3. As a suspension, with a finely divided silica powder such as Cab-O-sil.
4. In a gel, such as an aqueous, Triton X-100, mixture.

Quenching

Both coloured materials and certain chemicals can reduce the efficiency by either absorbing photons before they reach the photo-multiplier or reacting with the excited solvent molecules before they react with the scintillator. Problems exist in determining the efficiency. A number of techniques exist, these include:

(i) *Internal standard calibration.* The specimen of unknown activity A is counted (net count rate = C). It is then recounted (net count rate = $C + C_s$) after the addition of a known activity A_s of a non-quenching standard. The efficiency $E = C_s/A_s$, and the unknown activity $A = C/E = C \cdot A_s/C_s$.

(ii) *External standard calibration.* A series of samples of activity A, with various quenching factors, are counted (net count rate = EA) and then recounted (net count rate = $EA + CE$) after bringing an external gamma-ray source into a well-defined position near the sample. A calibration curve is thus obtained of the counting efficiency E versus the external count rate CE. Using this calibration curve, E for an unknown sample can now be determined from the observed CE.

(iii) *Channels-ratio method.* The counting channel is split into a lower part (a) and an upper part (b). The effect of quenching is to increase the net count rate (a, in channel a) and to decrease the net count rate (b, in channel b). The channels ratio (C_b/C_a) is thus a measure of Q, and since it depends only on the shape of the beta spectrum, it is independent of the sample activity. From a series of measurements on samples of known activity A, with different quenching factors, a calibration curve of channel-counting efficiency $E = (C_a + C_b)/A$ versus channels ratio (C_b/C_a) is obtained.

(iv) *External standard channels-ratio method.* This combines the two previous methods. Three channels are used, one set for the beta-emitter to be assayed, and the other two to determine the channels ratio (CE_b/CE_a) for the external standard. A set of samples of known activity are used to obtain a calibration curve of E versus (CE_b/CE_a).

Method (i) is most accurate, whereas method (iv) is recommended for low count rates or highly quenched samples because of statistical fluctuations in the sample count rate.

1.7. MEASUREMENT OF STOMATAL CONDUCTANCE AND PLANT WATER STATUS

by M. M. LUDLOW

1.7.1. Measurement of Stomatal Conductance

1.7.1.1. Introduction

In order to absorb CO_2 for photosynthesis plants expose to a dry atmosphere tissues which must be protected from dehydration. To overcome this dilemma, plants have evolved leaves with an epidermis composed of a relatively impermeable cuticle and turgor-operated valves—stomata. The epidermis not only reduces rates of CO_2 and water-vapour exchange, but it also provides a means of controlling assimilation and transpiration through the size of the stomatal pores. Thus stomata play a pivotal role in controlling the balance between water loss and biomass production. Measurement of the size of the stomatal opening (stomatal aperture) or of the resistance to CO_2 and water vapour (H_2O) transfer between the atmosphere and the internal tissues of the leaf imposed by the stomata (stomatal resistance) are important in many studies of biomass production.

1.7.1.2. Resistance or Conductance?

The Ohm's Law analogue of CO_2 and H_2O diffusion has been described in this volume by Long. The restriction to the movement of CO_2 and H_2O offered by the stomata is defined there as a *resistance* (conceptually similar to an electrical resistance). The size of the stomatal resistance is often compared with that of the boundary layer for water-vapour transfer, and with those of the boundary layer and the intracellular processes for CO_2 transfer. Comparison of resistances is both theoretically correct and biologically meaningful (Burrows and Milthorpe, 1976). However, if the limitation offered by stomata is being compared with the flux of CO_2 or H_2O, or being correlated with some biological or environmental variable such as leaf water status, it is more meaningful and less prone to misinterpretation to express it as a *conductance* (= 1/resistance) rather than as resistance (Burrows and Milthorpe, 1976; Hall *et al.*, 1976; Cowan, 1977). Fluxes are proportional to conductances but inversely proportional to resistances. However, most instruments are calibrated against physical resistances. Thus measurements are made as resistances and then conductances are calculated. Unfortunately this can lead to large errors when low resistances are converted to give high conductances because small systematic or random errors are relatively large when resistances are small.

1.7.1.3. Methods

Stomatal conductance can be obtained by determining the size of the stomatal aperture, or by measuring the rate of gaseous loss.

1.7.1.3(i). STOMATAL APERTURE

Stomatal aperture is usually measured by direct microscopic observation or by the extent or rate of infiltration of organic solvents (Kanemasu, 1975).

1.7.1.3(i)(a). *Direct Observation*

For a given leaf, plant or variety, the length and depth of stomata do not vary among stomata in mature tissues. Instead, most of the changes in aperture are associated with changes in width. In practice it is not possible to make a direct microscopic observation of stomata and at the same time preserve natural conditions. The change in conditions could alter the stomatal aperture. However, it is possible to make a stomatal impression before stomata have time to react by applying a quick-drying substance to the leaf surface (Kanemasu, 1975; Rice et al., 1979). The size of the stomatal aperture can be measured under a microscope either from the impression (where the stomatal pore is represented as a small raised area in what is equivalent to a photographic negative) or a positive which is made by painting the negative with a substance such as cosmetic nail varnish (stomatal pores appear as holes in the positive). Stomatal aperture can be converted into an equivalent diffusive resistance (or conductance) (Meidner and Mansfield, 1968; Meidner and Sheriff, 1976).

1.7.1.3(i)(b). *Infiltration by Liquids*

A series of mixtures of two liquids (0–100%) is made, one with a high, the other low, viscosity. The mixtures are then applied to leaf surfaces in sequence from the most to the least viscous. The first mixture to infiltrate the leaf is an index of the degree of stomatal opening. This index can be correlated for each species with aperture obtained by direct observation or with diffusive resistance (Kanemasu and Wiebe, 1975). The infiltration method is simple and cheap, but of limited accuracy.

1.7.1.3(ii). RATE OF WATER VAPOUR LOSS

Stomatal conductance can be calculated from rates of water-vapour loss. The most accurate way is by measuring water-vapour loss from leaves enclosed in leaf chambers using gas-exchange techniques (Davenport, 1975; see also Long in this volume). However, gas exchange, which is mainly a laboratory technique, is expensive and requires good technical support, especially if used in the field. Simpler, though less accurate, techniques are required for field

measurements where many determinations are usually required. In addition, instruments should be rugged, portable, battery-operated and relatively inexpensive (Kanemasu, 1975). Three such techniques will be described here.

1.7.1.3(ii)(a). *Cobalt Chloride Paper*

Paper impregnated with cobalt chloride is blue when dry and pink when moist. The time taken for a colour change when the paper is held against a leaf surface is an index of the rate of water loss and, hence stomatal conductance. The technique is quick, cheap, but only semi-quantitative. The time for the change in colour is linearly correlated with diffusive resistance measured with a diffusion porometer for sorghum ($r = 0.91$) and soybean ($r = 0.92$). The relationship may, however, vary with species and environmental conditions. Kanemasu and Wiebe (1975) describe the technique and give instructions for making cobalt chloride paper.

1.7.1.3(ii)(b). *Porometry*

There are basically two types of porometers. The mass-flow porometer measures the rate at which air is forced through (i.e. across) leaves under pressure (Hsiao and Fischer, 1975). The diffusion porometer measures the rate at which water vapour diffuses out of leaves. Mass-flow porometers are simple, cheap and usually do not involve electronic circuitry. Therefore they are useful in teaching, and for work at remote field sites. However, they have the following disadvantages: (i) they are best used for comparative rather than absolute measurement because of the errors and limitations; (ii) use is mainly restricted to leaves with stomates on both surfaces, although they can be used on hypostomatous leaves with special precautions; (iii) leakage of air from the apparatus, or more importantly, at the point of leaf attachment, can cause serious errors.

The resistance to mass flow can be correlated with stomatal aperture (Hsiao and Fischer, 1975) or with diffusive resistance for each species, depending upon stomatal distribution and the gas being considered. The viscous (or mass) flow resistance (Ω; $kg\,m^{-2}\,s^{-1}$) is given by

$$\Omega = \frac{\Delta P}{f_1}$$

where ΔP and f_1 are, respectively, the pressure gradient ($kg\,m\,s^{-2}$) and the flow of air across the leaf ($m^3\,s^{-1}$). A simple rough, generalization of the relationship between diffusive and viscous flow resistance to water vapour transfer is

$$r_s^{\mathrm{H_2O}} \propto \Omega^n$$

where n is estimated empirically to be 0.4.

Diffusion porometry is based on measurement of the rate of water-vapour loss from a leaf or portion of a leaf enclosed in a porometer chamber (Kanemasu, 1975). The rate of loss is determined from the rate of increase in humidity, or from the rate at which dry air is added to offset the increase in humidity due to transpiration (null-point porometer). In both approaches, water loss occurs from both the stomata and the cuticle. It is generally assumed that most of the loss occurs from the stomata, but the cuticular component becomes increasingly important as stomata close.

Those porometers which measure the rate of humidity increase have a humidity sensor in the chamber enclosing the leaf. Commonly used sensors are lithium chloride, sulphonated polystyrene, and solid-state thin-film semiconductors. The chamber can be non-aspirated (non-ventilated; Fig. 1.17(a)) or it can have a small fan which either stirs the air within the chamber (Fig. 1.17(b)) or forces air through a bypass

containing the sensor and back into the leaf chamber (Fig. 1.17(c)).

Aspirated porometers have a much lower boundary-layer resistance and this increases the sensitivity for measuring stomatal resistance. Moreover, the air in the chamber is well mixed. Both these characteristics are necessary to measure needle-like leaves or small branches which cannot be measured with the non-aspirated instruments. There are, to my knowledge, no commercial models of aspirated porometers.

Non-aspirated porometers are simple, relatively cheap, and are quite suitable for broadleaated plants. A number of commercial instruments are of this type: Lambda Autoporometer (Licor Ltd., Lincoln, Nebraska, U.S.A.); ΔT Automatic Porometer Mk II (Delta-T Devices, Cambridge, England); and Automatic Diffusion Porometer (Crump Scientific, Rayleigh, Essex, England).

Both aspirated and non-aspirated porometers are markedly affected by temperature. During measurements, the difference between leaf and sensor temperature should be $<1°C$ because isothermal conditions are usually assumed. In practice, this is achieved by shading the leaf for 1–2 minutes before measurement, and keeping the cup shaded during measurement. Moreover, porometers have to be calibrated at accurately controlled temperatures (better than $0.1°C$) and over a range of temperatures. They are calibrated against metal plates drilled with holes of known resistance. The porometer cup is clam-

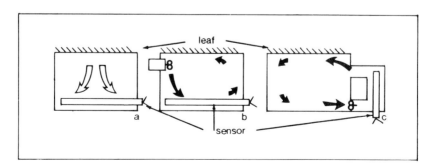

FIG. 1.17. Principles of operation of three types of diffusion porometer: (a) non-aspirated or non-ventilated, (b) ventilated by a fan within the chamber and (c) ventilated by pushing air through a bypass containing the sensor and back into the chamber.

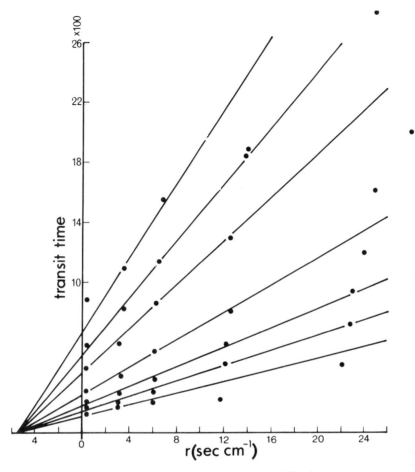

FIG. 1.18. Relationship between resistance (r) to diffusion of water vapour through holes on the calibration plate and time for relative humidity to increase over a predetermined range (transit time), determined at a range of temperatures. The point where the lines intersect is the cup resistance which is independent of temperature.

ped on one side of the plate and a wet blotting paper against the other side. The rate of increase of humidity at each temperature is measured for a series of combinations of size and number of holes. The time for humidity to increase over a particular range (e.g. 20–30% RH)—called the transit time—is plotted against the calculated physical resistance for each temperature (Fig. 1.18). A series of straight lines with a common intercept on the x-axis (which is the cup resistance) is obtained. The slopes of the lines are then plotted against temperature (Fig. 1.19). A polynomial can be fitted to this relationship and used to calculate the

slope for each measurement temperature. Alternatively, the slope can be obtained graphically and substituted in the following equation to calculate stomatal resistance (conductance):

$$r_s = \frac{\Delta t}{\text{slope}_t} - r_{\text{cup}}$$

where Δt is the transit times (in counts or seconds), slope_t is the slope for each measurement temperature and r_{cup} is the cup resistance determined from the relationship between transit time and calculated physical resistance (Fig. 1.18).

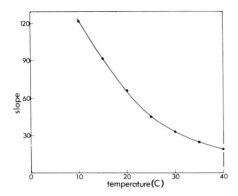

FIG. 1.19. Relationship between temperature and the slope of the line which describes the relation between resistance and transit time in Fig. 1.18.

Calibration is the most difficult part of diffusion porometry. The method based on physical resistance described above, or one based on the injection of known amounts of water vapour into the porometer cup can be used. Commercial manufacturers of diffusion porometers also provide both equipment for and details of calibration techniques.

In the null-point porometers, the rate at which dry air is added to the cup to balance the increase in humidity due to transpiration, or to keep humidity constant at the null point, is used to calculate stomatal resistance (Campbell, 1975).

If a flow rate of dry air F $(cm^3 s^{-1})$ is required to maintain a humidity h in a chamber enclosing a leaf of area A (cm^2), the rate of transpiration (T) would be:

$$T = Fe_a/A$$

where e_a = vapour pressure of the air in the chamber. At steady state, T can be written as the Ohm's Law analogue equation

$$T = (e_1 - e_a)/r$$

where e_1 is the saturated vapour pressure at leaf temperature and r is the resistance to water loss from the leaf.

Combining these two equations we get:

$$r = ((e_1/e_a) - 1)A/F$$

If the chamber is well stirred the boundary

layer resistance (r_a) is small. It can either be neglected, or determined by replacing the leaf with wet blotting paper or similar material (Landsberg and Ludlow, 1970). Moreover, if leaf and air temperatures are similar, e_1/e_a becomes $1/h$, the reciprocal of relative humidity. Thus stomatal resistance (r_s) is:

$$r_s = (((1/h) - 1)A/F) - r_a.$$

Calibration of the porometer involves calibration of flow rate rather than of resistance.

The advantages of null-point porometers are:

Humidity is constant during measurement and it is a steady state rather than a dynamic measurement. This could be important if stomata respond to humidity.

Measurements can be done at different humidities.

Calibration is more direct and less dependent on assumptions and uncertainties.

They can be used with needle-like leaves or small branches.

The disadvantage over other types of porometer are that they are more complex, they require more power and better technical support, and they are more expensive.

Bibliography and Further Reading

Burrows, F. J. and Milthorpe, F. L. (1976) Stomatal conductance in the control of gas exchange. In: *Water Deficits in Plant Growth*, Vol. 4 (Ed. T. T. Kozlowski), pp. 103–152. (Academic Press: New York.)

Campbell, G. S. (1975) Steady-state diffusion porometers. In: *Measurement of Stomatal Aperture and Diffusive Resistance* (Ed. E. T. Kanemasu), pp. 20–23, Bull. 809, College of Agric. Res. Center, Washington State University.

Cowan, I. R. (1977) Stomatal behaviour and environment. *Adv. Bot. Res.* **4**, 117–228.

Davenport, D. C. (1975) Stomatal resistance from cuvette transpiration measurements. In: *Measurement of Stomatal Aperture and Diffusive Resistance* (Ed. E. T. Kanemasu), pp. 12–15, Bull. 809, College of Agric. Res. Centre, Washington State University.

Hack, H. R. B. (1974) The selection of an infiltration technique for estimating the degree of stomatal opening in leaves of field crops in the Sudan and a discussion of the mechanism which controls the entry of test liquids. *Ann. Bot.* **38**, 93–114.

HALL, A. E., SCHULZE, E.-D. and LANGE, O. L. (1976) Current perspectives of steady state stomatal responses to environment. In: *Water and Plant Life* (Eds. O. L. LANGE, L. KAPPEN and E.-D. SCHULZE), pp. 169–188. (Springer-Verlag: Berlin.)

HSIAO, T. C. and FISCHER, R. A. (1975) Mass flow porometers. In: *Measurement of Stomatal Aperture and Diffusive Resistance* (Ed. E. T. KANEMASU), pp. 5–11, Bull. 809, College of Agric. Res. Center, Washington State University.

JARVIS, P. G. (1971) The estimation of resistances to carbon dioxide transfer. In: *Plant Photosynthetic Production a Manual of Methods* (Ed. Z. ŠESTÁK, J. ČATSKÝ and P. G. JARVIS), pp. 566–631. (Dr. W. Junk: The Hague.)

KANEMASU, E. T. (Ed.) (1975) *Measurement of Stomatal Aperture and Diffusive Resistance.* Bull. 809, College of Agric. Res. Center, Washington State University.

KANEMASU, E. T. and WIEBE, H. W. (1975) Other methods. In: *Measurement of Stomatal Aperture and Diffusive Resistance* (Ed. E. T. KANEMASU), pp. 23–24, Bull. 809, College of Agric. Res. Center, Washington State University.

LANDSBERG, J. J. and LUDLOW, M. M. (1970) A technique for determining resistance to mass transfer through the boundary layers of plants with complex structure. *J. Appl. Ecol.* **7**, 187–192.

MEIDNER, H. and MANSFIELD, T. A. (1968) *Physiology of Stomata.* (McGraw-Hill: London.)

MEIDNER, H. and SHERIFF, D. W. (1976) *Water and Plants.* (Blackie: Glasgow.)

MORROW, P. A. and SLATYER, R. O. (1971) Leaf resistance measurements with diffusion porometers: precautions in calibration and use. *Agric. Met.* **8**, 223–233.

RICE, J. S., GLENN, E. M. and QUISENBERRY, V. L. (1979) A rapid method for obtaining leaf impressions in grasses. *Agron. J.* **71**, 894–896.

STIGTER, C. J. (1972) Leaf diffusion resistance to water vapour and its direct measurement. I. Introduction and review concerning relevant factors and methods. *Meded Landbouwhogesch., Wageningen*, pp. 72–73.

1.7.2. Measurement of Plant Water Status

1.7.2.1. Introduction

The amount of water used directly in biochemical reactions is small compared with that transpired by plants and with the amount of water in plants at any one time. Plant water status strongly influences plant growth and biomass production, through its effect on leaf and root expansion and on photosynthesis. In general, biomass production is directly proportional to the supply and use of water. Therefore, measurement of plant water status is an important part of understanding biomass production and maximizing yield under irrigation.

Whereas it is generally accepted that water moves through the soil–plant–atmosphere system along gradients of water potential, there is still argument whether the water content or the water potential has the greater effect on physiological activity and on survival. Ideally, both content and potential should be measured in experiments, but other factors such as availability of equipment, and the nature of the experiment may dictate that only one of the two is measured.

1.7.2.2. Water Content

The amount of water in plant material can be expressed in a number of ways (Slavik, 1974):

$$\text{WATER CONTENT} = \frac{\text{fresh weight} - \text{dry weight}}{\text{dry weight}} \times 100\%.$$

RELATIVE WATER CONTENT

$$= \frac{\text{fresh weight} - \text{dry weight}}{\text{turgid weight} - \text{dry weight}} \times 100\%.$$

Turgid weight is obtained by floating leaves or leaf disks on water at the light-compensation point until constant weight is reached.

WATER SATURATION DEFICIT:

$$100 - \text{Relative Water Content}\%.$$

TURGID WEIGHT/DRY WEIGHT RATIO:

$$\frac{\text{turgid weight}}{\text{dry weight}}.$$

Determining water content requires relatively simple, cheap equipment and many samples can be taken as replicates or across a large number of treatments. However, the technique for determining relative water content requires considerable skill and precision to obtain accurate results. In addition, there are sometimes difficulties in interpreting results, because there is not much information on the relationship between relative water content and the rates of physiological processes.

There are also methods for non-destructive

measurement of water content such as using a β gauge, but these techniques are complex, expensive and require good technical support.

1.7.2.3. Water Potential

Water potential (Ψ is defined as the potential energy per unit mass of water with reference to pure water at zero potential)

$$\Psi = \frac{RT}{V_w} \ln a_w$$

where R is the gas constant, T is the absolute temperature, V_w is the partial molal volume of water and a_w is the activity of water. Water in most biological systems has less potential energy than pure water. This results in negative values for water potential.

The two main methods of measuring water potential are the pressure chamber technique and thermocouple hygrometry. Simpler, though less accurate, methods such as vapour equilibration and Shardakov's method (Slavik, 1974) are also available.

Thermocouple hygrometry techniques are based on placing soil or plant tissue into a small chamber, and allowing the water potential to come to equilibrium with the air in the chamber. The vapour pressure of this air is then measured by wet-bulb psychrometry or dewpoint hygrometry (Wiebe et al., 1971; Brown and van Haveren, 1972). These instruments are calibrated using blotting paper soaked in solutions of known osmotic potential. They are accurate but they are expensive, complex and require good technical support. Also, plant material may take many hours to reach equilibrium. This greatly reduces the number of measurements which can be made. These techniques are best used in the laboratory.

The pressure chamber is simple, cheap, rugged and ideally suited for field studies (Wiebe et al., 1971; Ritchie and Hinckley, 1975). A leaf cut from the plant is placed in the chamber with the cut end projecting through the hole in a rubber bung. The pressure applied to the leaf or branch to return the water interface to where it was before detachment, is equal and opposite to the tension in the xylem of the intact plant. Because

the osmotic potential of xylem sap is usually <0.2 kPa, the hydrostatic pressure in the xylem is equal to the water potential. The pressure-chamber technique has been reviewed extensively (Wiebe et al., 1971; Ritchie and Hinckley, 1975) and is now the most widely used method for characterizing plant water status.

Bibliography and Further Reading

BROWN, R. W. and VAN HAVEREN, B. P. (1972) Psychrometry in Water Relations Research. Utah Agricultural Experiment Station: Logan, Utah, U.S.A.

NOBEL, P. S. (1974) Biophysical Plant Physiology. (W. H. Freeman: San Francisco.)

RITCHIE, G. A. and HINCKLEY, T. M. (1975) The pressure chamber as an instrument for ecological research. Adv. Ecol. Res. 9, 165–254.

SLATYER, R. O. and McILROY, I. C. (1961) Practical Microclimatology. (UNESCO-CSIRO: Melbourne.)

SLAVIK, B. (1974) Methods of Studying Plant Water Relations. (Chapman & Hall: London.)

WIEBE, H. H., CAMPBELL, G. S., GARDNER, W. H., RAWLINS, S. L., CARY, J. W. and BROWN, R. W. (1971) Measurement of Plant and Soil Water Status. Bull. 484. Utah Agricultural Experiment Station, Logan, Utah, U.S.A.

Appendix 1.7. Practical Work on Stomatal Conductance

A1.7.1. Objective

To measure the stomatal conductance of leaves at various heights between the soil surface and the top of the crop and corresponding environmental factors (light, temperature, humidity and water potential) in order to determine which factor(s) is controlling stomatal conductance.

A1.7.2. Procedure

The following parameters should be measured.

(a) Photosynthetic Quantum Flux

Using the Lambda linear sensor and integrator and a metre rule, measure quantum flux above the crop and at a number of predetermined heights between the top of the crop and the soil surface. These measurements should be made at right angles to the rows, and from either the row or the centre of the inter-row area. If incident radiation is constant, measure above the crop and then at progressively lower heights to the soil surface. On the other hand, if incident radiation is changing, measure above the crop and then at a lower height, then above the crop again before measuring the next lowest position. The effect of changes in incident radiation can be compensated for by expressing the value at a particular height as a percentage of the incident value. This sequence of measurements should be repeated at five (5) different positions in the crop. The height of the crop should be measured at the same locations.

(b) Leaf and Air Temperature

Using the WESCOR thermocouple thermometer measure the temperature of the undersurface of a leaf at the top of the canopy. Then measure the air below the leaf, using the leaf to shade the thermocouple from the radiation. Repeat these measurements at a number of predetermined heights between the top of the crop and the soil surface, and at five (5) different locations. Also measure soil temperature by carefully pushing the thermocouple into the soil to a depth of 1 cm. Wipe the thermocouple clean after it is removed from the soil.

(c) Wet- and Dry-bulb Temperature

Using the Assman psychrometer measure the wet- and dry-bulb temperature, above and at a number of heights in the plant canopy, and at five (5) different locations in the crop.

If possible visit the meteorological station and read the wet- and dry-bulb temperature on the unaspirated psychrometer. If not, make arrangements to obtain this data.

(d) Leaf Water Potential

Moisten a piece of paper towel and place it around the leaf before cutting the leaf with a razor blade. Take the leaf quickly to the pressure chamber and determine the balance pressure. Do this at various positions in the canopy and at five (5) locations in the crop.

(e) Short-wave and Visible Radiation

Use one radiometer to measure short-wave radiation and the other, which is fitted with a filter to eliminate the visible wavelengths, 400 to 700 nm, to measure the non-visible component of short-wave radiation. The difference between these two readings gives a value for visible radiation. Measure with both instruments both above the crop and at various positions in the canopy and five (5) locations in the crop. Place the radiometers parallel to the rows and as close to the plants as possible.

(f) Stomatal Conductance

Shade the leaf for 1 minute before placing the porometer cup so that the stomatal conductance of the lower surface is measured. Keep the porometer cup shaded during measurement. Allow the porometer to go through several cycles until a constant value of the transit time is obtained. Read leaf temperature before removing the porometer cup. Make measurements at various heights in the canopy and at five (5) locations in the crop.

A1.7.3. Calculations
The following calculations should be made.

(a) Photosynthetic Quantum Flux (mE m^{-2} s^{-1})

Express the quantum flux at each level as a percentage of the incident value. Determine a mean value for each height for each of the five locations. Plot relative quantum flux against height above ground.

(b) Leaf and Air Temperature (°C)

Plot mean leaf and air temperature for each height at the five locations against height above the ground.

(c) Wet- and Dry-bulb Temperature (°C)

Using the slide rule provided, calculate saturation deficit (kPa) from wet- and dry-bulb temperature. Details of this calculation are given on page 15 and below. Plot mean saturation deficit and mean dry-bulb temperature for the five locations against height above the ground (1 kPa = 10 mbar).

(d) Leaf Water Potential (MPa)

Calculate leaf water potential in MPa (= 10 bar) and plot mean leaf water potential for the five locations against height above the ground.

(e) Short-wave and Visible Radiation (W m^{-2})

Calculate short-wave and non-visible radiation from readings and the calibration data supplied. Subtract these values to get visible radiation. Calculate the proportion of non-visible radiation in short-wave radiation. Plot both mean visible radiation (W m^{-2}) and the mean proportion of non-visible for the five locations against height above the soil surface.

(f) Stomatal Conductance (cm s^{-1})

From the transit time and the temperature, calculate stomatal resistance (r_s) using the formula:

$$r_s = \frac{\text{transit time}}{\text{slope}_{\text{temp}}} - r_{\text{cup}} \quad (\text{s cm}^{-1})$$

where r_{cup} is the cup resistance and slope$_{\text{temp}}$ is the slope of the transit time–resistance relationship at the measurement temperature. Relationship between transit time and resistance, and between slope and temperature for the porometer will be provided.

Calculate stomatal conductance (= 1/stomatal resistance).

Plot mean stomatal conductance for the five locations against height above the soil surface.

A1.7.4. Analysis of Data
(a) Each student should compare the profiles (parameter plotted against height) of environmental factors and stomatal conductance. Then plot stomatal conductance against the factor(s) which correlate best with it. Write one (1) paragraph giving your

interpretation of which factor(s) is controlling stomatal conductance.
(b) Answer the following questions:
  1. Does the proportion of non-visible radiation change with depth in the canopy? If so why does it change?
  2. Does the difference between leaf and air temperature change with height and why?
  3. How could biological factors such as leaf position and leaf age complicate your interpretation of factors controlling stomatal conductance (in two sentences)?
  4. Identify possible errors in the measurements and briefly say how they could be improved.

A1.7.5. To Calculate Saturation Deficit from Wet- and Dry-bulb Temperatures (Using Casella Slide Rule)
  1. Set cursor over wet bulb temp. (T_w) on scale 1.
  2. Bring wet-bulb depression (dry bulb temp. − wet bulb temp.; $T_d - T_w$) on scale 2 (for non-aspirated) or scale 2B (for aspirated) beneath the cursor.
  3. Read dew-point temperature on scale 1 against the zero of scale 2 (or 2B in red, aspirated).
  4. Set RIH on scale 3 against 100 on scale 4.
  5. Set cursor over dew-point temperature (determined in 3) (on scale 3).
  6. Read vapour pressure (e) on scale 4.
  7. With RIH still on scale 3 against 100, on scale 4 set cursor over dry-bulb temperature (T_d) and read saturated vapour pressure (e_0) on scale 4.
  8. Saturation deficit (w_e) = saturated vapour pressure − vapour pressure = $e_0 - e$ (mbar).
  9. Units: 1 mbar = 0.75 mm Hg @ 0°C
                    = 0.1 kPa.

Appendix 1.8. Results of a 24-hour Experiment

The following figures (Fig. 1.20, Figs. 1–6) illustrate a set of results obtained by the students attending the U.N.E.P. Photosynthesis

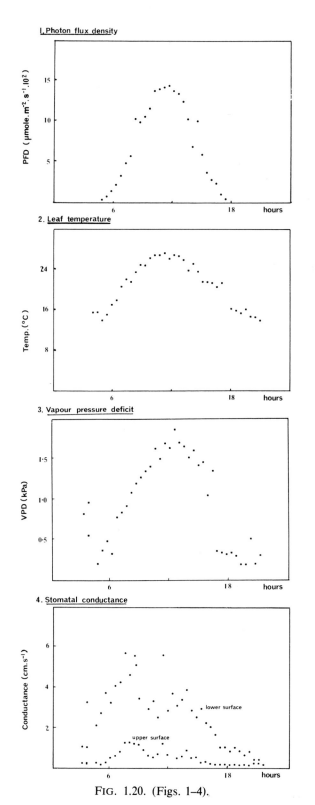

FIG. 1.20. (Figs. 1–4).

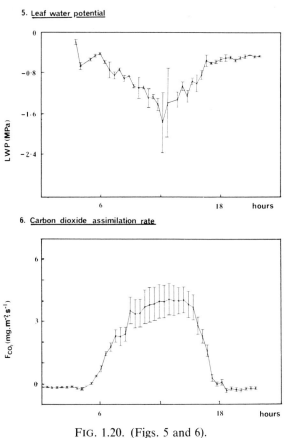

FIG. 1.20. (Figs. 5 and 6).

and Bioproductivity Training Course held at the Maize Research Institute in Belgrade. The students worked in shifts so that a full 24 hours of data for one day could be obtained. The following day was used for analysis of the data and production of the figures illustrated.

The experiment was conducted on a 3-week-old crop of *Phaseolus vulgaris* which had reached the five-leaf stage. The investigation was limited to the most recently expanded leaf, which was the third true leaf in this crop.

Figure 1 illustrates changes in photon flux density measured with a quantum sensor through the 24-hour period.

Figures 2 and 3 illustrate changes in the leaf temperature and the atmospheric vapour-pressure deficit measured with thermocouples and an Assman psychrometer respectively.

Figure 4 illustrates changes in stomatal

conductance (g_s) for both the lower (higher set of points) and upper (lower set of points) leaf surfaces.

Figure 5 illustrates change in leaf water potential ($\pm$ standard error) measured with a pressure chamber.

Figure 6 illustrates the rate of CO_2 uptake per unit leaf area per unit time, measured on five replicate leaves using a gas sample selector and an infra-red gas analyser. Points indicate mean values ($\pm$ standard errors).

Integration of the measured rates of CO_2 uptake suggested a net assimilation over the 24-hour period of $12.9\,\mathrm{g\,m^{-2}}$, giving an efficiency of conversion of mols of photons of 0.9%.

A1.9. General Appendices

A1.9.1. Terms Used in Plant-Growth Analysis (see Section 1.4 for practical measurement)

Term	Definition	Dimensions
AGR (Absolute Growth Rate)	Change in dry wt. of a plant per unit time	dW/dt (weight time^{-1})
C (Crop Growth Rate)	Change in dry wt. of a crop per unit area of ground per unit time	$\dfrac{1}{P}\dfrac{dW}{dt}$ (weight area^{-1} time^{-1})
L_a (Leaf Area)	The total leaf area of a plant or other unit of vegetation at one point in time	L_a (area)
L (Leaf Area Index)	The leaf area of a crop above a unit area of ground	L_a/P (dimensionless)
P (Sample Area)	An area of ground defined with a crop	P (area)
R (Relative Growth Rate)	Change in dry wt. of a plant or crop per unit of dry wt. per unit of time	$\dfrac{1}{W}\dfrac{dW}{dt}$ (weight weight^{-1} time^{-1})
t	Time	t (time)
E (Unit Leaf Rate)	Change in dry wt. of a plant or crop per unit of leaf area per unit of time	$\dfrac{1}{L}\dfrac{dW}{dt}$ (weight area^{-1} time^{-1})
W	Weight of plant or crop at a given point in time	W (weight)

A1.9.2. Appendix of Symbols and Definitions of Terms in Photosynthetic Gas Exchange and Plant Microclimate

Symbol	Subscripts	Definition	Units
A		Projected leaf area	$\mathrm{m^2}$
C		Concentration of CO_2 in air. Mass of CO_2 per unit volume of air	$\mathrm{mg\,m^{-3}}$
	C_a	In the ambient air	
	C_w	At the leaf internal air space/ mesophyll interface	
D		Diffusion coefficient of a gas in air	

A1.9.2. (*cont.*)

Symbol	Subscripts	Definition	Units
e		Partial pressure of water vapour in air	kPa
	e_a	In the ambient air	
	e_s	Saturation vapour pressure for a given temperature	
f		Rate of air flow	$m^3\,s^{-1}$
F		Net flux of gas across a unit area of leaf	$mg\,m^{-2}\,s^{-1}$
	F_{CO_2}	CO_2 flux = (gross photosynthetic rate–respiratory rate) also P_n	
	F_{H_2O}	Water-vapour flux = transpiration rate	
I		Irradiance, flux of radiant energy into a unit area	$W\,m^{-2}$
I_p		Photon flux density, flux of visible photons into a unit area	$mol\,(photons)\,m^{-2}\,s^{-1}$
r^{CO_2}		Resistance of a diffusion pathway to CO_2 flux (or water-vapour flux $-r^{H_2O}$)	$s\,m^{-1}$
	r_g	Gas phase resistance (equals sum of boundary layer and stomatal resistance)	
	r_m	Liquid phase resistance (also termed residual, intracellular of mesophyll resistance)	
T		Temperature of a body	°C
	T_a	Air temperature	
	T_l	Leaf temperature	
VPD		Leaf vapour-pressure deficit. The difference of partial pressures of water vapour in the ambient air and in air saturated with water vapour at the temperature of the leaf, i.e. $e_s - e_a$	kPa
Δ		As a prefix; the difference between measurement of a quantity	
Γ		The CO_2 compensation point of photosynthesis. The concentration of CO_2 in air at which F_{CO_2} is zero, for an individual leaf or plant	$mg\,m^{-3}$
ψ		Water potential. The chemical potential of a body of water relative to pure water divided by the molal volume	MPa
	ψ_l	Leaf water potential	
	ψ_e	Soil water potential	
χ		The absolute humidity of air. Mass of water per unit volume (subscripts as for C)	$mg\,m^{-3}$

NET LEAF PHOTOSYNTHESIS

$mg\,CO_2\,dm^{-2}\,h^{-1}\,0.0278 = mg\,CO_2\,m^{-2}\,s^{-1}$

$mg\,CO_2\,dm^{-2}\,h^{-1}\,0.6312 = \mu mol\,CO_2\,m^{-2}\,s^{-1}$

$mg\,CO_2\,m^{-2}\,s^{-1}\,22.72 = \mu mol\,CO_2\,m^{-2}\,s^{-1}$

$mg\,CO_2\,dm^{-2}\,s^{-1}\,22.72 = \mu mol\,CO_2\,dm^{-2}\,s^{-1}$

CO_2 CONCENTRATION (NTP)

$ppm\,CO_2\,1.964 = mg\,CO_2\,m^{-3}$

$ppm\,CO_2\,44.65 = \mu mol\,CO_2\,m^{-3}$

$mg\,CO_2\,m^{-3}\,22.72 = \mu mol\,CO_2\,m^{-3}$

A1.9.3. Light Intensity

	Erg cm^{-2} s^{-1}	mwatt cm^{-2}	W m^{-2} = joule m^{-2} s^{-1}	cal cm^{-2} s^{-1}	cal cm^{-2} min^{-1}
Erg cm^{-2} s^{-1}	1	10^{-4}	10^{-3}	2.389×10^{-8}	1.434×10^{-6}
mwatt cm^{-2}	10^4	1	10	2.389×10^{-4}	1.434×10^{-2}
Joule m^{-2} s^{-1} = watt m^{-2}	10^3	0.1	1	2.389×10^{-5}	1.434×10^{-3}
cal cm^{-2} s^{-1}	4.186×10^7	4.186×10^3	4.186×10^4	1	60
cal cm^{-2} min^{-1}	6.976×10^5	69.76	697.6	0.01667	1

1 watt = 1 joule s^{-1}, 1 joule = 0.2389 cal.

Day light, full sun: 950 W m^{-2} = 1.36 cal cm^{-2} min^{-1} ≈ 95,000 lux
  from these PAR (400–700 nm) is: 1800 μmol photons m^{-2} s^{-1} ≈
  ≈ 399 W m^{-2} = 0.572 cal cm^{-2} min^{-1} = 42% of total. 1 W m^{-2} (total) ≈ 1.895 μmol m^{-2} s^{-1}.

Blue sky light: 72 W m^{-2} = 0.103 cal cm^{-2} min^{-1} ≈ 9000 lux
  from these PAR is: 200 μmol photons m^{-2} s^{-1} ≈
  ≈ 45 W m^{-2} = 0.065 cal cm^{-2} min^{-1} = 63%. 1 W m^{-2} (total) ≈ 2.778 μmol m^{-2} s^{-1}.

Hg high-pressure lamp 400 W: 153 W m^{-2} = 0.219 cal cm^{-2} min^{-1} ≈ 24,000 lux
  from these PAR is: 350 μmol photons m^{-2} s^{-1} ≈
  ≈ 78 W m^{-2} = 0.111 cal cm^{-2} min^{-1} = 52%. 1 W m^{-2} (total) ≈ 1.370 μmol m^{-2} s^{-1}.

Conversion Table for Radiometric and Photometric Energy Units

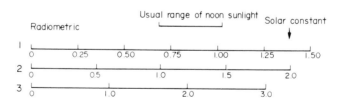

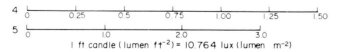

1 ft candle (lumen ft^{-2}) = 10.764 lux (lumen m^{-2})

1. $\times 10^3$ watts m^{-2}, $\times 10^6$ erg cm^{-2} s^{-1}, $\times 10^3$ joules cm^{-2} s^{-1}
2. $\times 10^0$ cal cm^{-2} min^{-1}
3. $\times 10^{-2}$ cal cm^{-2} s^{-1}
4. $\times 10^4$ ft candle from clear sky.
5. $\times 10^4$ ft candle from tungsten lamp.

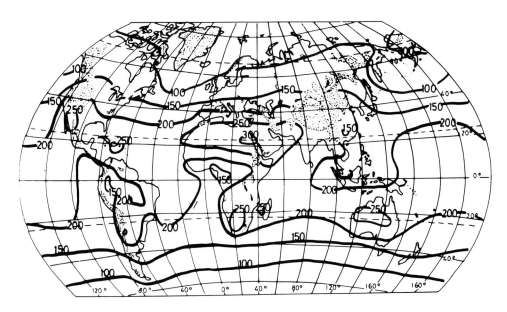

FIG. 1.21. Annual mean global irradiance on a horizontal plane at the surface of the earth (W/m^2 averaged over 24 hours).

SHOOT MORPHOLOGY AND LEAF ANATOMY IN RELATION TO PHOTOSYNTHETIC EFFICIENCY

by H. R. BOLHÀR-NORDENKAMPF

2.1. SHOOT MORPHOLOGY AND THE RELATIONSHIP OF SINGLE LEAF TO WHOLE PLANT CO₂ ASSIMILATION

Leaves of different age are attached to the shoot at different heights in annual C_3 crops. Each leaf represents a stage of ontogenetic development. Therefore an ontogenetic gradient may be observed from the top of the growing plant to its base. Microclimatic factors change significantly with height in the canopy. Thus, crop plants have an optimum canopy density for high biomass productivity. This is attained when the resulting microclimate in the canopy corresponds with the different physiological properties of the leaves situated at different heights.

A developing young leaf of the canopy grows in full sun light. Some weeks later the same leaf is totally expanded and ontogenetically mature, but it is now shaded by the newly developed younger leaves and receives only 20% of full sunlight or even less. This change in the light conditions of the canopy induces the young leaves to function as high-light or "sun leaves" and the older ones to function as low-light or "shade leaves". This microclimatic situation seems to be taken into consideration by the normal ontogenetic development of leaves.

Normally the youngest leaf of an annual dicotyledon grows in full sunlight until it is fully expanded and all the important physiological properties are well developed. These leaves have high photosynthetic rates at high light intensities.

As there is much leaf area in this stage of development their contribution to biomass production of the whole canopy is unexpectedly low. The highest proportion of productivity comes from the lower leaves which are shaded for a part of the day. These "semi-shade" leaves inside the canopy show relatively low photosynthetic rates at higher light intensities but have a good efficiency at low light as is typical of many C_3 plants. If data for photosynthetic measurements are used to estimate the productivity of growing canopies only data from whole plants or even from parts of the whole canopy should be considered. The rates of net leaf photosynthesis of a distinct leaf can be used in such calculations only if its ontogenetic and physiological properties with respect to position in the canopy are well known. In each level of the canopy the rates of net leaf photosynthesis have to be determined. If the leaf area in each zone is estimated the productivity of the canopy can be roughly calculated. The major problem of such measurements in well-growing canopies is the continuous physiological change. Nevertheless, in a canopy with completed vegetative development and with known stable physiological zones you can obtain a good idea of the overall productivity of the canopy by measuring rates of net leaf photosynthesis of a representative sample of leaves. Comparisons of data on net photosynthesis is only possible when the leaves used are well defined in age, position and physiological state (Fig. 2.1).

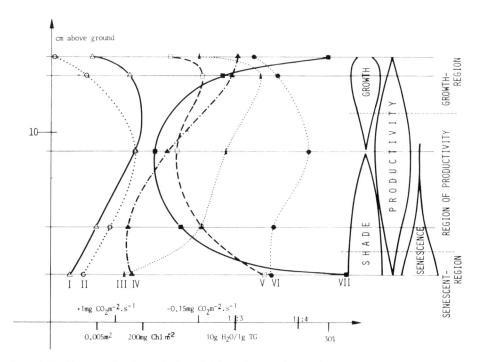

FIG. 2.1. Characterization of the physiological regions of a bean plant, *Phaseolus vulgaris* var. *nanus* L. by several parameters.

   I △————△ Net leaf photosynthesis in mg CO_2 m^{-2} s^{-1} at 25°C, 700 mg CO_2 m^{-3} and 350 W m^{-2}.

   II ○········○ Leaf area in m^2.

  III ▲·········▲ Chlorophyll content in mg m^{-2}.

  IV ▲—·—▲ Dark respiration in mg CO_2 m^{-2} s^{-1}.

   V □————□ Relative water content in g H_2O per g dry weight.

  VI ●·····● Chlorophyll a to b relationship.

 VII ■—■ Dark respiration as percent of net leaf photosynthesis.

2.2. LEAF ANATOMY

Leaves are protected by a cutinized epidermis. Gas exchange with the atmosphere is largely limited to stomatal pores, in the epidermis, surrounded by the guard cells which are able to open and close the aperture. The leaf parenchyma, the mesophyll, is in many cases divisible into a palisade parenchyma and a spongy parenchyma. The elongated cells of the palisade parenchyma are exposed to the intercellular air only in part. The spongy parenchyma consists of cells of irregular shape which build a wide system of intercellular spaces (see Fig. 2.2). Embedded in the mesophyll are the vascular bundles, commonly called veins. In the smaller vascular bundles of the leaf it can be clearly

seen that the conductive tissue is enclosed by one or more layers of compactly arranged cells, forming the bundle sheath(s). The proportions of these four components (epidermis, mesophyll, bundles and intercellular spaces) varies between species. Within a species these proportions are modified by climatic and ontogenetic factors. Great differences in leaf anatomy are found between "sun leaves" and "shade leaves" (see Table 2.1 and Fig. 2.2).

2.2.1. Leaf Anatomy and Light

The anatomy of shade leaves together with their physiological properties enable such leaves to use low-light intensities very efficiently by an increase in the number of reaction centres and

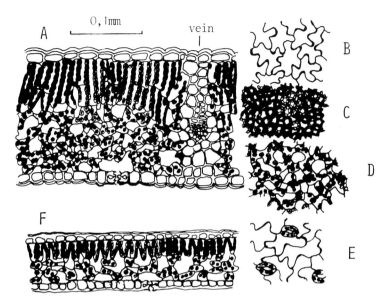

FIG. 2.2. A, Cross-section "sun-leaf" of a typical mesophyte. B, Horizontal section upper epidermis. C, Horizontal section palisade mesophyll. D, Horizontal section spongy mesophyll. E, Horizontal section lower epidermis. F, Cross-section "shade-leaf".

Table 2.1. *Typical changes in leaf anatomy according to the light environment*

	High-light sun leaves	Low-light shade leaves
Cuticle	+	−
Leaf thickness	+	−
Palisade parenchyma	up to 4 cell layers	less than 1 cell layer
Spongy parenchyma	+	−
Intercellular spaces	−	+
Stomata per mm^2 leaf area	240	130
Stomata size	−	+
Chlorophyll content per leaf area	+	−
Chlorophyll content per plastid	−	+
Grana size	−	+
Chlorophyll a:b relation	up to 1:4	less than 1:3
Ribulose bis-phosphate carboxylase activity	+	−
Glycolate oxidase activity	+	−
P_{700} per cytochrome f, per ferredoxin or per plastoquinone	−	+

associated antennae systems feeding to the components of the electron transport chain. This is indicated by changes in the ratio of P_{680} and P_{700} to cytochrome f and ferredoxin (Table 2.1). However, one result of this is that "shade leaves" cannot make use of high-light intensities, the plastids of such leaves may even be destroyed by photooxidative processes in high light. So one function of the multi-layer palisade parenchyma in high-light leaves may be as protection against high-light intensities. In this case the spongy parenchyma with its extensive leaf air system may be the principal photosynthetic tissue. High-light leaves have many small stomata which are able to regulate transpiration more precisely and so CO_2 uptake remains possible even when little water is available (Larcher, 1975).

2.2.2. Xeromorphy

Of similar interest are the morphological and anatomical adaptations to dry habitats. Here, the living and transpiring tissue is reduced and surrounded by dead sclerenchyma cells. The stomata are sunken on the upper side of the leaf and the leaf blade is often involuted. A thick cuticle, trichomes and sunken stomata increase the boundary-layer resistance to gas exchange and reduce the transpiration rate. Plants which have to withstand dry conditions for short periods only may develop an epidermal water-storage tissue. Such water-storage tissues together with a good protection against transpiration losses and a special photosynthetic pathway (CAM), which causes stomatal opening only at night, enables succulent plants to survive extreme drought conditions over many months.

2.2.3. Leaf Anatomy in the Gramineae

The C_3 gramineae have two bundle sheaths: an inner thick-walled "mesotome" sheath without chloroplasts surrounded by a second parenchyma sheath with a few chloroplasts sometimes smaller than those of the mesophyll. The bundle-sheath cells do not play a significant role in photosynthetic CO_2 assimilation and metabolism. CO_2 is fixed into the mesophyll cells by ribulose-1,5-bis-phosphate carboxylase and starch is formed in the same chloroplasts.

In the C_4 gramineae a single- or a double-bundle sheath as in C_4 dicotyledons is present. The bundle sheath cells are very large relative to the

mesophyll and contain specialized chloroplasts which differ from mesophyll plastids by being larger, synthesizing starch in a normal photoperiod and often by reduced size of grana. Around this bundle-sheath chlorenchyma the mesophyll is arranged radially so that each mesophyll cell is in direct contact with a bundle-sheath cell or is no more than one cell removed. This arrangement of the chlorenchyma is termed the "Kranz" syndrome. This anatomy is essential to the functioning of the C_4 pathway. The CO_2 is fixed by phosphoenol pyruvate (PEP) carboxylase in the mesophyll cytoplasm into oxalacetate which is reduced to malate or transaminated to aspartate. These substances are transported into the bundle-sheath cells. Here CO_2 is generated by a decarboxylation process and refixed by the ribulose-1,5-bis-phosphate (RuBP) carboxylase. The previous fixation by PEP carboxylase in the mesophyll cells probably functions as a mechanism for concentrating CO_2 in bundle-sheath cells to inhibit RuBP oxygenation. This allows RuBP carboxylase to work efficiently even with low CO_2 concentrations in the intercellular spaces of the leaf. This combination of anatomic and metabolic features adapts C_4 plants to hot and dry habitats with high irradiation.

2.2.4. Sub-groupings of C_4 Species

Species (grasses: eu-panicoid, andropogonoids, C_4 dicots) with a mesotome sheath only are malate formers. In the mesophyll cells the first product of CO_2 fixation, oxalacetate, is converted predominantly to malate. The malate is transported to the bundle-sheath cells and is decarboxylated by NADP-dependent malic enzyme. This reaction gives rise to CO_2, $NADPH_2$ (reducing power) and pyruvate. By the generation of $NADPH_2$ in the decarboxylation process no additional photoreduction in the bundle-sheath plastids seems to be necessary and so these plastids have no or poorly developed grana and are poor in PS II. The big chloroplasts in the sheath cells are centrifugal in position in grasses and centripetal in position in dicots. This category of C_4 species

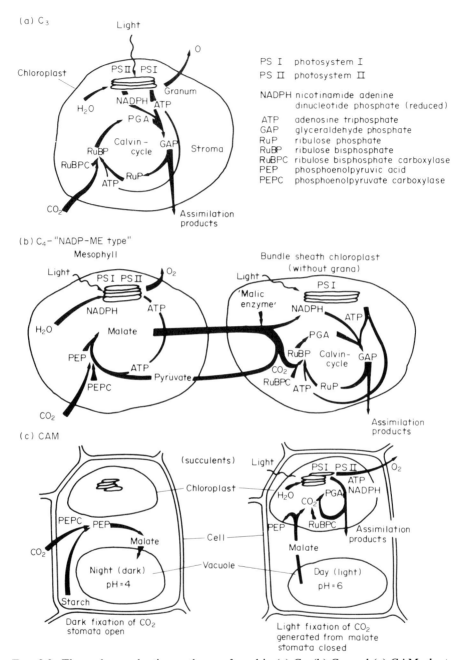

FIG. 2.3. The carbon reduction pathways found in (a) C_3, (b) C_4, and (c) CAM plants.

is termed the "NADP-ME type" after the major decarboxylase of the sheath tissue.

All species (grasses: chloridoid, ergastoid, C_4 dicots) with two bundle sheaths are aspartate formers. If aspartate is transported to the bundle-sheath cells practically no reducing power to reduce the CO_2 is generated in the

decarboxylation process. Here the sheath chloroplasts develop grana and PS II activity. Two subtypes can be distinguished in this group:

The sheath chloroplasts are irregular in size and are centrifugal in position. After the transamination of aspartate to oxalacetate CO_2 is

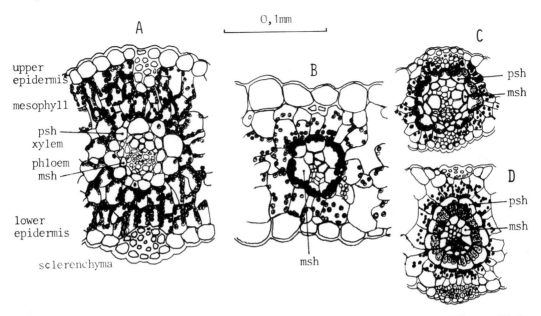

FIG. 2.4. Cross-sections of C₃ and C₄ leaves. (A) C₃ species, (B) C₄ species "NADP-ME" type, (C) C₄ species "PCK" type, (D) C₄ species "NAD-ME" type, psh = parenchyma bundle sheath, msh = mesotome bundle sheath.

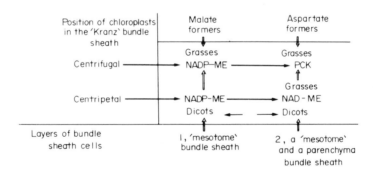

FIG. 2.5. Scheme of correlation between anatomy and physiology in C₄ plants.

generated by PEP-carboxykinase in the sheath cytoplasm. These are the "PCK-type" C₄ species.

In the second group the sheath chloroplasts are centripetal in position. The aspartate from the mesophyll cells is converted first to oxalacetate and secondly to malate to be decarboxylated by the NAD-malic enzyme in the sheath mitochondrica. Though reducing power arises in the last step, this "NAD-ME" grass species and C₄ dicots have grana and PS II activity in the sheath chloroplasts.

The relationships between the anatomy and photosynthetic carbon reduction pathways of plants are summarized in Figs. 2.3, 2.4, and 2.5.

2.3. EXPERIMENTS

2.3.1. Leaf Anatomy

Transverse sections of the leaf blade cut by hand, using a razor blade, are examined in a light microscope. The layers of the palisade parenchyma cells are counted and the relation to the thickness of the palisade parenchyma to the spongy parenchyma can be estimated. Look for schlerenchyma cells and examine the leaf-air system in the different parts of the mesophyll. Do this by drawing a part of the cross-section.

Try to find the stomata and describe their position and arrangement: raised, sunken, in crypts, protected by trichomes or wax; hypostomatic, epistomatic or amphistomatic. To count the stomata per unit of leaf area you have to use a good incident light microscope. You may count the stomata on artificial replicas if leaves have a blade without many trichomes and if the stomata are not deeply sunken. To produce the replicas you can use any transparent paint. The best is a water-soluble acrylic paint or varnish. Put a drop of the varnish on the surface of the leaf and pull off the dry transparent film with forceps after some minutes. Now you can count the stomatal impressions on this replica by the help of a calibrated grid in the ocular of the light microscope. Take replicas of different parts of the leaf because stomatal frequency varies greatly on one leaf. Less than 60 stomata per mm^2 and more than 600 stomata per mm^2 indicate that the plant grows in an extreme habitat (xeromorphic, hygromorphic respectively). Do not measure stomatal widths by the technique of replicas or epidermal strips. It is better to use an incident light microscope with a mirror (reflecting) objective for such measurements. Alternatively stomatal aperture may be determined by indirect methods as detailed in Section 1.7.

2.3.2. Stomatal Width

An estimate of the relative stomatal aperture may be obtained by the infiltration method using xylol, alcohol (ethanol) and paraffin oil. Put a drop of each solution on the upper and lower surface of the leaf and judge the degree of infiltration by the size of the area darkened by the liquid entering the internal air spaces. The results may give good response curves of the changes of the climatic factors over one day of one plant only, but you *cannot* compare the response curves of different plants. Table 2.2

Table 2.2

Time	*Substance		
	1	2	3
Sun rise	•	•	
8^h	⊗	⊖	
10^h	◖	◉	•
12^h	◖	·	•
14^h	⊖	·	
16^h	◖	⊖	•
18^h	⊗	•	
Sun set	⊖	•	

*1 = xylol; 2 = alcohol; 3 = paraffin oil

illustrates the results typically obtained during the course of a day with a mature leaf of *Phaseolus vulgaris*.

2.3.3. Demonstration of the O$_2$ Evolution with Whole Plants

Fill a large glass container with a 0.01% indigo carmine (indigo disulphonate) solution. While stirring constantly, add carefully single drops of a 10% sodium dithionite solution until the blue indigo carmine is reduced to the yellow form. Place a whole plant or a branch into this solution and seal the container, excluding air. After several minutes of illumination, if there is no excess sodium dithionite, blue areas will appear around all green parts of the plant. The reason is that the small amounts of O$_2$ produced by photosynthesis inside the leaf are enough to reoxidize the yellow indigo carmine back to the blue form.

2.3.4. *In situ* Demonstration of Photosystem II Activity

The presence of PS II activity (reducing power) can be demonstrated by using the reduction of tetranitro-blue–tetrazolium chloride. Cross-sections of the leaf blade, which need not be very thin, are cut by hand and infiltrated under vacuum with the staining solution (see below). After 5 to 20 minutes of illumination, under a microscope for instance, the areas in the leaf transverse section containing chloroplasts with high PS II activity will show a dark-blue colour. This occurs first in all wounded cells, then in the mesophyll cells of all plants and finally in the granal bundle-sheath plastids of the C_4 grasses of the NAD-ME and PCK groups. The agranal bundle sheath plastids of the NADP-ME species turn blue only after several hours of illumination. Misleading results may occur if infiltration is poor because of residual air in the intercellular spaces. Note that the bundle-sheath cells which are longitudinally elongated are more likely to be injured during sectioning than the more rounded mesophyll cells. Thus C_4 plants, especially dicotyledons, may show some coloured bundle-sheath chloroplasts even after a short period of illumination.

Stock solutions:
(A) 0.1% tetranitro blue terazolium chloride (1 mg per ml). (Do not make up more than 10 ml at one time; store dye powder and solution in refrigerator.)
(B) 0.1 M phosphate buffer, pH 6.0.
(C) 0.3 M sucrose.
Staining solution: 1 part A, 3 parts B, 1 part C.

See Downton *et al.* (1970).

2.3.5. Differentiation Between C_3 and C_4 Plants by Detection of Starch *in situ*

Cross-sections of the blades of leaves of various types are stained by immersion in iodine solution. Starch will have been formed in the chloroplasts if the leaf has been illuminated for a long time and net leaf photosynthesis has been proportionally high as compared to translocation of photosynthate. In such leaves chloroplasts will be stained dark blue by iodine. This occurs in chloroplasts of the mesophyll cells of all C_3 plants. In C_4 plants such starch accumulation occurs preferentially in the "Kranz" bundle-sheath cells. The detection of starch in this way is more distinct if chlorophyll

is extracted from the leaves, using hot alcohol, before staining.

After 2 days in the dark the chloroplasts in the leaf are generally destarched. Such a leaf can be used to produce high resolution starch prints or pictures. A photographic negative with high contrast is mounted on the upper surface of the leaf exposed to light. Starch will be formed only in those areas which are actually reached by light rays. The starch formation will be proportional to incident light intensity. Thus if the leaf is cut off, killed and extracted in hot water, followed by hot alcohol (on a water bath), and stained with iodine the negative will be reproduced as a positive image. The high resolution image obtained indicates that starch is in general only formed in those cells which are actually illuminated and that translocation of photosynthate from an illuminated chloroplast, to a non-illuminated chloroplast, even when they both lie in the same cell, does not readily occur.

Solution for iodine stain (Lugol). First disolve 2 g KI in 5 ml water, then add 1 g I and make up to 300 ml with water.

Bibliography and Further Reading
DOWNTON, W. J. S., BERRY, J. A. and TREGUNNA E. B. (1970) C_4-photosynthesis: Non-cyclic electron flow and grana development in bundle sheath chloroplasts. *Zeitschrift für Pflanzenphysiologie* **63**, 194–198.
ELLIS, R. P. (1977) Distribution of the Kranz syndrome in the southern African ergrostoideae and panicoideae according to bundle sheath anatomy and cytology. *Agroplantae* **9**, 73–110. (Botanical Research Institute Pretoria, 0002.)
ESAU, K. (1977) *Anatomy of Seed Plants*, 2nd edition. (J. Wiley & Sons: New York/Santa Barbara/London/Sydney/Toronto.)
LARCHER, W. (1975) *Physiological Plant Ecology*. (Springer-Verlag: Berlin, Heidelberg and New York.)
MOLISCH, H. (1972) *Populäre biologische Vorträge*. (Fischer: Jena).
WALKER, D. A. and ROBINSON, S. P. (1978) Chloroplast and cell. A contemporary view of the photosynthetic carbon assimilation. *Berichte der deutschen botanischen Gesellschaft* **91**, 513–526.
WILD, A. (1979) Physiology of photosynthesis in higher plants. The adaptation to light intensity and light quality (in German). *Berichte der deutschen botanischen Gesellschaft* **92**, 341–364.

ALGAE: GROWTH TECHNIQUES AND BIOMASS PRODUCTION

by A. VONSHAK and H. MASKE

3.1. INTRODUCTION

The importance of algal culture in physiological and biochemical research and the fact that algae are one of the most efficient converters of solar energy to useful energy have increased the interest in algal-culturing techniques. Algal-culturing techniques can be divided into two categories: the first applies to laboratory conditions and a controlled environment and the second to outdoor conditions for the larger-scale production of biomass.

This section aims to provide the student with the basic techniques and concepts of algal culture and to present a state of the art report on algal biomass production, its problems and achievements. It is not our intention to provide a complete comprehensive manual of the techniques of algal cultivation, and the student is encouraged to make use of the bibliography at the end of this section. We will, however, try to present an overall approach to algal-growth techniques and to point out problems that the scientists should be aware of when planning his own growth system and choosing the parameters to be measured.

Algae are photosynthetic, non-vascular plants that contain at least one pigment—chlorophyll a. This general definition does not reflect the diversity of size and form of this group of organisms. In this section, we discuss only aquatic *alga* which grow photoautotrophically and reproduce solely by cell-division.

3.2. GROWTH OF MICRO-ALGAE: TECHNIQUES AND KINETICS

The algae dealt with in this section behave as simple microorganisms undergoing a simple non-sexual life cycle multiplying only by cell division. Hence common growing techniques and mathematical analysis used in bacteriological studies may also be applied to such algal cultures. Algae may be grown in batch cultures or in continuous culture.

3.2.1. Batch Culture

A batch culture is initiated by the transfer of a small portion of a culture into a new culture medium, resulting in growth and an increase in biomass. Biomass concentration can be measured in many ways, as cell number, dry weight, packed cell volume, or in terms of any convenient biochemical component or parameter. The rate of increase in biomass concentration is generally expressed by the specific growth rate (μ), which is calculated according to the following formula:

$$\mu = \frac{dx}{dt} \cdot \frac{1}{X}$$

where X = biomass concentration; i.e. μ = increase in biomass per unit time per average biomass, the units are t^{-1}.

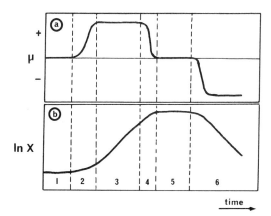

FIG. 3.1. Schematic representation of changes in (a) growth rate (μ) and (b) biomass (X) as a function of time in batch culture. The numbers refer to the various growth phases described in the text.

The changes in specific growth rate during the development of a culture is shown in Fig. 3.1(a). Figure 3.1(b) shows the increase of biomass concentration (X) with time (t) in such a batch culture. These figures indicate that the following phases can be distinguished in the growth of a batch culture: (1) lag phase, (2) accelerating phase, (3) logarithmic phase (balanced growth), (4) decelerating phase, (5) stationary phase, (6) death phase. Each growth phase is a reflection of a particular metabolic state of the cell population at any given time. These phases of growth are considered in more detail below.

1. *Lag phase.* A newly transferred culture may have a lag phase for several reasons:

(a) The population transferred may have been in a metabolically "bad" ("shifted down") state. This case occurs when the inoculum is taken from the stationary or death phase of the parent culture.

(b) The freshly inoculated batch culture has first to become conditioned to the culture medium (e.g. through the chelation of metals by excretion products).

(c) The measured biomass parameter does not take the non-viable portion of the population into account, and therefore, the biomass production of the small but vigorously growing portion of viable cells is masked by the non-viable cells representing the major part of the population.

2. *Acceleration phase.* In this phase different biomass parameters increase sequentially: the first component to increase is RNA, followed by protein and then dry weight. Cell number is usually the last parameter to show an increase. This phase may also be referred to as "shift up".

3. *Logarithmic phase.* During this period the growth rate remains constant and the biomass concentration changes according to the formula:

$$X_{t_2} = X_{t_1}\, e^{\mu(t_2-t_1)}.$$

Thus

$$\mu = \frac{\ln x_2 - \ln x_1}{t_2 - t_1}.$$

The mean doubling time (dt) or generation time (g) thus can be determined as follows:

$$g = \frac{\ln 2}{\mu} = \frac{0.693}{\mu}.$$

During this phase of growth the concentration ratio of the different biochemical components stays constant as shown in Fig. 3.2. This pattern of growth is termed "balanced growth", it may also be referred to as exponential growth.

4. *Deceleration phase.* During this phase the biochemical composition changes in a sequence opposite to that in the acceleration phase. This may also be termed "shift down".

5. *Stationary phase.* During this phase the biomass parameter remains constant. Other parameters may increase or decrease. The final concentration of biomass reached at the stationary phase is usually taken to be a function of the depletion of either some essential nutrient in the medium which becomes limiting or due to limitations resulting from gas exchange. Other factors that might determine the biomass concentration at the stationary phase include excretion products that inhibit further growth, possibly due to changes in pH of the medium, or shading in dense algae cultures.

6. *Death phase.* A decline in biomass concentration occurs as a result of an increase in

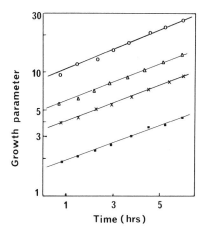

FIG. 3.2. Increase in several growth parameters with time in a culture of *A. nidulans* during exponential or balanced growth: (O—O) = chlorophyll, $(\mu g\, ml^{-1}) \times 10$; (△—△) = dry weight, $(\mu g\, ml^{-1}) \times 10^{-1}$; (■—■) = cell number, $ml^{-1} \times 10^{-7}$; (× — ×) = optical density, $A_{560\,nm} \times 10$.

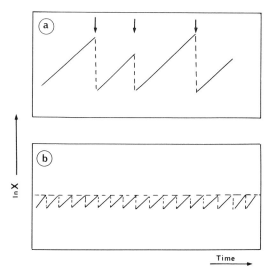

FIG. 3.3. Growth patterns in (a) semicontinuous and (b) turbidostat cultures. The arrows in (a) represent the points at which the culture was diluted. The dotted line in (b) represents the preset turbidity value. X = the biomass concentration measured as optical density.

the ratio of respiration to photosynthesis above one, or as a result of cell death and lysis.

3.2.2. Continuous Culture

1. *Semicontinuous cultures.* Semicontinuous cultures are a type of batch cultures which are diluted at frequent intervals (Fig. 3.3(a)). The biomass concentration in these cultures has to be monitored so that the frequency of dilution and the dilution ratio (volume of culture: volume of new medium) can be estimated.

2. *Turbidostat culture.* This type of culture (Fig. 3.4) is similar in nature to a semicontinuous culture, except that the monitoring of the biomass is not performed manually but by an optical device that controls the dilution rate maintaining the culture at a preset optical density (Fig. 3.3(b)). In this case turbidity is used as the measure of biomass concentration. In such turbidostat cultures growth is such that nutrients are always non-limiting. Light, on the other hand, may be a limiting factor. A schematic representation of the control cycle for a turbidostat culture is shown in Fig. 3.5(a).

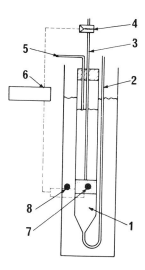

FIG. 3.4. Schematic diagram of a turbidostat: (1) culture vessel; (2) air + CO_2 inlet; (3) medium inlet; (4) magnetic valve; (5) air and medium outlet; (6) electronic control device; (7, 8) measuring and reference photo cells.

a) <u>Turbidostat</u>

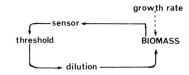

b) <u>Chemostat</u>

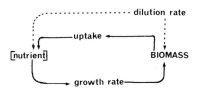

FIG. 3.5. The control cycles in (a) a turbidostat and (b) a chemostat.

3.3. CHEMOSTAT CULTURES

The principle of chemostat culture is based on the relationship between the specific growth rate and a limiting nutrient concentration that regulates the growth rate in such a way that it matches a preset constant dilution rate. The chemostat represents the most simple of the continuous culture devices, because the regulatory mechanism is dependent on the metabolism of the organism. The chemostat is the only culture method from which data can be obtained on the relationship between growth rate and nutrient limitations of the organisms. The control cycle for a chemostat is shown in Fig. 3.5(b). The chemostat system (Fig. 3.6) consists of a reservoir containing the culture medium which has a constant concentration of the limi-

ting nutrient; a constant-flow device (pump, capillary with constant hydrostatic pressure) and a culture vessel with constant volume (maintained by an overflow). Culture fluid leaves the vessel at the same rate as the new medium is fed into the vessel by the flow device. The culture has to be homogeneously mixed in the culture vessel. With algae this mixing may be achieved by the aeration stream, or mechanical means (using an impeller or, with a smaller culture vessel, a magnetic stirrer).

In both chemostats and turbidostats the rate of production of cells through growth is equal to the rate of loss of cells through the overflow. Hence, when the flow rate (pump rate) $f = dv/dt$ (where v is the culture volume) and the dilution rate $D = f/v$, then at steady state:

$$\mu \cdot X = \frac{dv(X)}{dt \cdot v} = D \cdot X$$

hence

$$\mu = D.$$

3.4. SYNCHRONIZED CULTURES

The three techniques described above are used for studying the growth pattern of populations of microorganisms which contain cells at

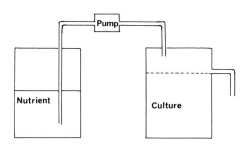

FIG. 3.6. Schematic diagram of a chemostat.

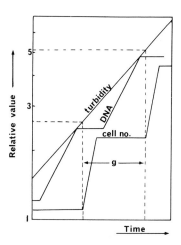

FIG. 3.7. The changes in some growth parameters in a synchronized culture.

all stages of the life cycle. In contrast, synchronized cultures are used for studies of individual growth patterns of cells in culture, since in this type of culture the cells are all at the same stage of the cell cycle at any given time. The pattern of growth and the changes in some of the growth parameters are shown in Fig. 3.7. Synchronized cultures can be obtained using either of the following techniques: (i) induction of synchrony by repeated shifts in the environmental or nutritional conditions; (ii) physical separation of cells and subsequent reculturing of those that are at the same stage of the cell cycle, from a random population.

3.5. NUTRIENT-LIMITED GROWTH

Figure 3.8 illustrates some hypothetical results from a chemostat culture showing the relationships between biomass concentration (as determined by measuring the amount of limiting nutrient taken up), the limiting nutrient concentration and the production of biomass as a function of the dilution rate (observed growth rate/maximum growth rate). The production of biomass (which equals flow rate times biomass concentration—the units being weight of biomass produced per time interval) exhibits a clear maximum at a growth rate close to the maximal growth rate. This picture neglects the dependence of the relative yield (biomass production per unit of nutrient used) on growth rate. The ratios of biomass parameters (other than those associated with the limiting nutrient) to the limiting nutrient incorporated into the organism changes in characteristic ways at different growth rates. For substrates providing energy (organic nutrients such as sugar, etc.) the amount of biomass produced per unit of nutrient increases with increasing growth rate (Fig. 3.9). This is a result of a change in the ratio of anabolism to catabolism. If the limiting nutrient is an inorganic salt or an essential vitamin the amount of biomass produced per unit of limiting nutrient decreases with increasing growth rate (Fig. 3.10).

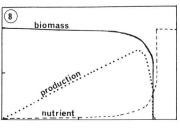

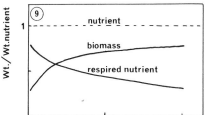

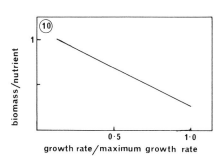

FIG. 3.8. Schematic representation of changes in biomass concentration, nutrient concentration and production as a function of dilution rate.

FIG. 3.9. The relationship between biomass production or loss of energy substrate through respiration as a function of the fractional growth rate.

FIG. 3.10. The production of biomass on a nutrient basis (yield) for a "non-energy" nutrient as a function of the fractional growth rate.

3.6. ALGAL BIOMASS PRODUCTION

During the past 30 years considerable attention has been paid to the possibilities offered by large-scale culturing of micro-alga as a biomass source of single-cell protein. The potential advantages of such systems, e.g. their high photosynthetic efficiency, high growth rate (doubling times of 3 to 10 hours), high yields (20 to $30 \, g \, d^{-1} \, m^{-2}$) using the entire plant with its high protein content (50 to 70% w/w), and the possibility of using brackish or saline water are balanced out by the disadvantages of high costs of production and initial investment. The high production costs are a result of both biological and technological problems that have yet to be solved. These will be considered in terms of those problems associated with the biology of algal biomass production and those problems associated with the technical and engineering aspects.

3.6.1. Biological Problems

1. *Water source.* Algae growing in brackish or saline water in ponds are often faced with the problem of the increasing salinity caused by evaporation. Thus, when the water source is brackish or saline the organism of choice will be one that is capable of growing successfully in a wide range of saline conditions and thus of adapting rapidly to changes in salinity.

2. *Optimization of growth.* It is quite clear that the economics of algal biomass production under outdoor conditions can be improved by increasing the amount of biomass obtained per unit area. This aim can be achieved by optimizing growth conditions. While the nutritional requirements of the algae culture can easily be met by an excess supply of nutrients, environmental limitations, such as light and temperature, are more difficult to control in outdoor systems. The productivity of the system, in terms of dry weight produced per unit time per unit area, is a function of both the growth rate and the algae concentration, while the availability of solar radiation to a single cell in the culture is a function of the total radiation and the cell concentration. In a culture in which there are no other environmental limitations, the maximal yield will depend on the optimal cell concentration, so regulation of cell concentration is thus of the utmost importance. The optimal cell concentration is a function of the algal species, the total radiation, the degree of induced turbulence and other environmental

limitations, and is in particular influenced by the depth of the algal culture.

3. *Contamination.* In open systems, contamination by other algae species may take place and, therefore, the product may not be uniform. This problem can sometimes be solved by choosing growth conditions that give a selective advantage to the required algal species.

4. *Choice of the algae species.* Although there are about 20,000 species of alga, only a few of them have been studied under outdoor conditions. The possibility of cultivating other algal species out of doors and their potential use for the production of biomass should be investigated. The organism of choice should exhibit the following characteristics:

   (i) fast growth rate;
   (ii) wide range of tolerance to temperature and radiation extremes;
   (iii) high content of protein, lipids or carbohydrates, or selectively accumulate a specific metabolite (e.g. glycerol);
   (iv) ease of harvesting.

Other requirements may be added to this list depending on the specific approach and requirements of the system.

3.6.2. Engineering Problems

Pond construction. Since the first experiments on outdoor algal biomass production, which took place in the early 1950s, much progress has been made in the field of pond construction and operation. One of the main components of the total capital investment is the cost of constructing a large-scale algal production unit. At present, the technology developed for surface irrigation has been applied to the manufacture of algal ponds in the form of wide channels 6–8 m wide and 20–30 cm deep. The channels are covered with black PVC or polyethylene and turbulence is produced with a paddle wheel.

Separation and dehydration. These two factors seem to be the major component of the high energy requirement in the processing of algal biomass and, therefore, any improvement in the process will increase the economic feasibility of the overall algal biomass production. The use of filtration or flocculation instead of centrifugation for separation will decrease the energy input. However, any of these mechanical techniques will yield a paste containing only 15 to 20% dry matter. Hence, the energy must also be used for drying. The development of a drying system using solar energy seems a promising approach.

Uses of algal biomass. Today it seems that the idea of using algal biomass as a protein source for general human consumption is not practically feasible. It is, however, thought that biomass production systems should be developed for such diverse purposes as: wastewater treatment; production of fish food; production of commercial chemicals, such as glycerol, mannitol or lipids; production of biochemicals and extraction of pigments such as natural dyes; and production of energy by fermentation of the biomass.

Reviewing the past 30 years of research on algal biomass production, we feel that a regional approach, i.e. the exploitation of local resources to solve local problems, seems to be more promising than the application of algal biomass production to the solution of global problems.

3.7. WHERE TO OBTAIN ALGAL CULTURES

1. The Culture Collection of Algae, Department of Botany, Indiana University, Bloomington, Indiana, U.S.A.
2. The Culture Centre of Algae and Protozoa, Storey's Way, Cambridge, England.
3. The Culture Collection of Algae and Microorganisms, Institute of Applied Microbiology, University of Tokyo, Bunkyo-ku, Tokyo, Japan.
4. Universitaet Goettingen, Botanisches Institut, Algae Sammlung, Goettingen, Federal Republic of Germany.

Bibliography and Further Reading

CARR, N. G. and WHITTON, B. A. (1973) *The Biology of Blue-green Algae*, (Blackwell.)

FOGG, G. E. (1975) *Algal Cultures and Phytoplankton Ecology.* (The University of Wisconsin Press: Madison.)

LEWIN, R. A. (ed.) (1962) *Physiology and Biochemistry of Algae.* (Academic Press.)

MEGNELL, G. G. and MEGNELL, J. E. (1970) *Experimental Bacteriology.* (Cambridge University Press.)

PIRT, S. J. (1975) *Principles of Microbe and Cell Cultivation.* (Blackwell.)

STANIR, K. Y. (1976) *The Microbial World*, 4th Edition. (Prentice-Hall Inc.)

STEWART, W. D. P. (1974) *Algal Physiology and Biochemistry.* (University of California Press: Berkeley.)

Handbook of Phycological Methods,
    Vol. 1. *Culture Methods and Growth Measurements*, 1973, edited by J. R. STEIN.
    Vol. 2. *Physiological and Biochemical Methods*, 1978, edited by J. A. HELLEBUST and J. S. CRAIGIE.
    Vol. 3. *Cytological and Developmental Methods*, 1979, edited by E. GANTT.
    All published by Cambridge University Press.

Algal Culture: From Laboratory to Pilot Plant, edited by J. S. BURLEW. (Pub. Carnegie Inst. of Washington, 1953.)

SHELEF, G., SOEDER, C. J. and BALABAN, M. (eds.) (1980) *Algal Biomass.* (Elsevier: North-Holland.)

APPENDIX

A.3.1. Occurrence of Pigments in the Various Classes of Algae

	Chlorophyll					Phyco-bilin	Carotene		Xanthophyll	
	a	b	c	d	e		α	β	Lutein	Fuco-xanthin
Cyanophyceae	+	−	−	−	−	+	−	+	(+)	−
Euglenophyceae	+	+	−	−	−	−	(+)	+	+	−
Pyrrhophyceae	+	−	+	−	−	+	+	+	−	−
Chrysophyceae	+	−	+	?	?	−	+	+	?	+
Xanthophyceae	+	−	+	−	+	−	−	+	−	−
Chlorophyceae	+	+	−	−	−	−	+	+	+	−
Phaeophyceae	+	−	+	−	−	−	−	+	+	+
Rhodophyceae	+	−	−	+	−	+	+	+	+	−

A.3.2. Useful Definitions on Algal Cultures

D = dilution rate = the flow rate $(f) = dv/dt$ divided by the culture volume.

D_c = critical dilution rate—when the specific growth rate is at a maximum and any increase in the flow rate will result in a wash-out of the culture.

g = mean generation time and in a continuous culture equals the retention time. The retention time (t_d) in a homogeneously mixed culture is on the average equal to 0.693 (ln 2) divided by the specific growth rate (μ).

μ = specific growth rate = $dx/dt.X$.

Y = growth yield = the increase in biomass resulting from consumption of a given amount of substrate = dx/ds.

Productivity = the rate of production of biomass in terms of dry weight per unit of time per volume.

A.3.3. Technical Problems

The user of chemostat or turbidostat cultures should be aware of the following technical problems:

Chemostats

1. Non-exponential washout due to bad mix-

Turbidostats

1. Wall growth as above, or condensation within the light path.

2. Fluctuations in the performance of flow devices.

General

1. Contamination of air supply, particularly by oil drops if a pressurized air supply is used. This problem can be overcome by use of suitable filters.

A.3.4. Analytical Techniques

Cell counting. In determining cell-division rate or correlating the increase in one of the cell components to cell number, the need for cell counting is quite obvious. The following table describes the most common commercial cell-counting devices available. They are used mainly in laboratory cultures. For counting natural diluted populations the use of an inverted microscope and settling chambers is recommended. An alternative is to concentrate the algae and refer to the table for easy counting of cell concentration.

Cell-counting Devices

Commercial name	Chamber volume (in ml)	Depth (mm)	Objectives for magnification	Cell size (mm)	Cell concentration counted easily
Redgwick Rafter	1.0	1.0	2.5–10	50–100	$30–10^7$
Palmer Malony	0.1	0.4	10–45	5–150	$10^2–10^5$
Speirs Levy hemacytometer	$10^{-3} \times 4$	0.2	10–20	5–75	$10^4–10^7$
Improved Neaubouer	$10^{-4} \times 2$	0.1	20–40 (phase)	2–30	$10^5–10^7$
Petroff Houser	2×10^{-5}	0.02	40–100 (phase)	0.5–5	$10^5–10^8$

ing or growth of high density of cells on solid supports within the vessel, e.g. walls, stirrers.

2. Variation in the flow rate due to perishability of tubing in the peristaltic pumps or formation of deposits in cylinder pumps or fluctuations in temperature and hydrostatic pressure in the capillaries.

3. Inhomogeneous biomass due to differentiation into sexual or resting stages.

Dry weight. For direct estimation of biomass concentration or for measurement of biomass production the most common technique used is determination of dry weight. Although it is a simple method, certain steps should be taken carefully.

1. *Sampling.* Insure that stirring of algal suspension is adequate and that pipetting is

done quickly to prevent algal settling during the process of sampling.

2. *Washing.* Washing of samples is important to insure the removal of salt and other particles (care should be taken to avoid osmotic shock, in particular with marine algae or algae grown on media of high osmotic potential).
3. *Drying.* Samples should be dried to a constant weight.
4. *Cooling.* Samples must be cooled in dry air before being weighed.

The procedure for dry weight determination is as follows:

1. Separation of algal cells from water by centrifugation or by filtration through fibreglass or a nitrocellulose filter.
2. Drying of samples to a constant weight. (Too short a period of drying will result in incomplete removal of water; too long a period may cause oxidation in the dry material.)

Expression of results. The results of dry-weight determination are expressed per unit of volume or in outdoor ponds per illuminated area of algal culture.

Chlorophyll determination. The Unesco method as detailed below is generally used in phytoplankton studies. In physiological studies some modifications are made:

1. centrifugation instead of filtration;
2. using hot methanol for extraction of green algae and 80% acetone for blue-greens.

The Unesco method is as follows. (This may differ in detail from methods described elsewhere in this manual. However, adoption of this standard method by people working in the field of algology enables direct comparison of results from various laboratories.)

This procedure was agreed to by a SCOR/UN-ESCO working group and has been published by Unesco in *Monographs on Oceanographic Methodology.*

METHOD

CONCENTRATION OF SAMPLE

Use a volume (Note *a*) of sea water which contains about 1 μg chlorophyll *a*. Filter (Note *b*) through a filter (Note *c*) covered by a layer of $MgCO_3$ (Note *d*).

STORAGE

The filter can be stored in the dark over silica gel at 1°C or less for 2 months but it is preferable to extract the damp filter immediately and make the spectrophotometric measurement without delay.

EXTRACTION

Fold the filter (plankton inside) and place it in a small (5–15 ml) glass, pestle-type homogenizer. Add 2–3 ml 90% acetone. Grind 1 min at about 500 rpm. Transfer to a centrifuge tube and wash the pestle and homogenizer 2 or 3 times with 90% acetone so that the total volume is 5–10 ml. Keep 10 min in the dark at room temperature. Centrifuge (Note *e*) for 10 min at 4000–5000 *g* (Note *f*). Carefully pour into a graduated tube so the precipitate is not disturbed and if necessary dilute (Note *g*) to a convenient volume (Note *h*).

MEASUREMENT

Use a spectrophotometer with a band-width of 30 Å or less, and cells with a light path of 4–10 cm (Note *i*). Read the extinction (optical density, absorbance) at 7500 (Note *j*), 6630, 6450, and 6300 Å against a 90% acetone blank.

CALCULATION

Subtract the extinction at 7500 Å from the extinctions at 6630, 6450, and 6300 Å. Divide the answers by the light path of the cells in centimetres. If these corrected extinctions are E_{6630}, E_{6450}, and E_{6300} the concentrations of chlorophylls in the 90% acetone extract as μg/ml are given by the SCOR/UNESCO equations. If the values are multiplied by the volume of the extract in millilitres and divided by the volume of the seawater sample in litres, the concentration of the chlorophylls in the sea water is obtained as μg/litre (= mg/m^3).

NOTES

(a) The amount of chlorophyll *a* should be less than 10 μg, otherwise a second extraction with 90% acetone might be necessary. With ocean water about 4–5 litres of sample should be used; with coastal and bay waters, sometimes one-tenth of this amount is sufficient.

(b) Use no more than two-thirds of full vacuum.

(c) Satisfactory filters include paper (Albet), cellulose (Cella "grob"), and cellulose ester (0.45–0.65 μm pore-size); the filter should be 30–60 mm in diameter. If these filters clog with inorganic detritus, use Schleicher & Schüll 575.

(d) Add about 10 mg $MgCO_3/cm^2$ filter surface, either as a powder or as a suspension in filtered sea water.

(e) A swing-out centrifuge gives better separation than an angle centrifuge.

(f) If a stoppered, graduated centrifuge tube is used, the extract can be made up to volume and the supernatant carefully poured or pipetted into the spectrophotometer cell.

(g) If turbid, try to clear by adding a little 100% acetone or distilled water or by centrifuging again.

(h) This depends on the spectrophotometer cell used. The volume should be read to 0.1 ml.

(i) Dilute with 90% acetone if the extinction is greater than 0.8.

(j) If the 7500 Å reading is greater than 0.005/cm light path, reduce the turbidity as in Note g.

$$C_a = 11.6\ E_{665} - 1.31\ E_{645} - 0.14\ E_{630}$$
$$C_b = 20.7\ E_{645} - 4.34\ E_{665} - 4.42\ E_{630}$$
$$C_c = 55.0\ E_{630} - 4.64\ E_{665} - 16.3\ E_{645}.$$

Turbidity. Turbidity is often used as a rapid and reliable parameter for determination of algal growth. The optical extinction of a well-mixed culture is determined at a wavelength between 450 and 650 nm, or in a colorimeter using a yellow or green filter. Care should be taken to ensure that the algae are not clumped (gentle homogenization may be necessary) or that they sediment during the measurement.

A.3.5. Culture Media

If cultures are ordered ask for the details of the media used for the given culture at the centre. Some useful media are as follows:

Allens's Medium

For each 1000 ml of medium required, add the following to 966 ml of glass-distilled water:

		Stock g/200 ml H_2O
$NaNO_3$	1.5 g	—
K_2HPO_4	5 ml	1.5
$MgSO_4 . 7H_2O$	5 ml	1.5
Na_2CO_3	5 ml	0.8
$CaCl_2$	10 ml	0.5
$Na_2SiO_3 . H_2O$	10 ml	1.16
Citric acid	1 ml	1.2
Trace elements*	1 ml	

*Detailed below.

Adjust pH to 7.8. Solidify, if desired, with agar (10 g/l).

Medium for *Anacystis nidulans*

Stock solutions	g/l	Medium ml/l
1 M $MgSO_4 . 7H_2O$	246.5	0.5
1 M $CaCl_2 . 2H_2O$	147.0	0.1
4 M $NaCl$	233.7	0.5
0.04 M $Na_2MoO_4 . 2H_2O$	10.0	0.1
2 M KNO_3	202.2	10.0
1 M K_2HPO_4	174.2	12.0*
Metals solution A_5	—	1.0
Fe-EDTA	—	1.0

Metals solution A_5	g/l
H_3BO_3	2.86
$MnCl_2 . 4H_2O$	1.81
$Na_2MoO_4 . 2H_2O$	0.252
$ZnSO_4 . 7H_2O$	0.222
$CuSO_4 . 5H_2O$	0.079

Fe-EDTA solution

Dissolve 16 g EDTA (free acid) and 10.4 g KOH in 186 ml dist. water. Mix this solution with other containing 13.7 g $FeSO_4 . 7H_2O$ (with low Mn content) in 364 ml dist. water. Bubble air through the mixture to oxidize Fe^{++} to Fe^{+++} (3–4 h). Final pH should be about 3.

Medium for *Chlamydomonas reinhardtii*

	Minimal medium (g)	High salt minimal medium (g)
NH_4Cl	0.05	0.50
$MgSO_4 . 7H_2O$	0.02	0.02
$CaCl_2 . 2H_2O$	0.01	0.01
K_2HPO_4	0.72	1.44
KH_2PO_4	0.36	0.72
Hutner's trace elements (see next page)	1 ml	1 ml
Distilled water	1 litre	1 litre

*K_2HPO_4 solution. Place 12 ml of 1 M K_2HPO_4, to which 0.84 g $NaHCO_3$ had been recently added, in a separate tube. Sterilize (20 min, 1 atm) and leave to cool. Mix both solutions under aseptic conditions.

Hutner's trace elements solution

	(g)
EDTA	50.0
$ZnSO_4 . 7H_2O$	22.0
H_3BO_3	11.4
$MnCl_2 . 4H_2O$	5.1
$FeSO_4 . 7H_2O$	5.0
$COCl_2 . 6H_2O$	1.6
$CuSO_4 . 5H_2O$	1.6
$(NH_4)_6Mo_7O_{24} . 4H_2O$	1.1
Distilled water	750 ml

Boil, cool slightly, and bring to pH 6.5–6.8 with KOH (do not use NaOH). The clear solution is diluted to 1000 ml with distilled water and should have a green color which changes to purple on standing. It is stable for at least one year.

For heterotrophic acetate mutants the media may be supplemented with sodium acetate at a concentration in the medium of 0.20%.

Medium for *Chlorella*

	g/l
KNO_3	1.25
KH_2PO_4	1.25
$MgSO_4 . 7H_2O$	1.00
$CaCl_2$	0.0835
H_3BO_3	0.1142
$FeSO_4 . 7H_2O$	0.0498
$ZnSO_4 . 7H_2O$	0.0882
$MnCl_2 . 4H_2O$	0.0144
MoO_3	0.0071
$CuSO_4 . 5H_2O$	0.0157
$Co(NO_3)_2 . 6H_2O$	0.0049
EDTA	0.5

The pH of the medium is 6.8.

Medium for *Euglena*

To 1000 ml of pyrex-distilled water add:

Sodium acetate	1.0 g
Beef extract	1.0 g
Tryptone	2.0 g
Yeast extract	2.0 g
$CaCl_2$	0.01 g

If desired, may be solidified by adding 15 g agar.

Medium for *Spirulina*

	g/l
NaCl	1.0
$MgSO_4 . 7H_2O$	0.2
$CaCl_2$	0.04
$FeSO_4 . 7H_2O$	0.01
EDTA	0.08
K_2HPO_4	0.5
$NaNO_3$	2.5
K_2SO_4	1.0
$NaHCO_3$	16.8

and 1 ml/l of A_5 and B_6 as below:

A_5	g/l
H_3BO_3	2.86
$MnCl_2 \cdot 4H_2O$	1.81
$ZnSO_4 . 7H_2O$	0.222
$CuSO_4 . 5H_2O$	0.074
MoO_3	0.015

B_6	g/l
NH_4NO_3	229.6×10^{-4}
$K_2Cr_2(SO_4)_4 . 24H_2O$	960×10^{-4}
$NiSO_4 . 7H_2O$	478.5×10^{-4}
$Na_2SO_4 . 2H_2O$	179.4×10^{-4}
$Ti(SO_4)_3$	400×10^{-4}
$Co(NO_3)_2 . 6H_2O$	439.8×10^{-4}

Sea water media (f/2)

Nutrients	g/l
$NaNO_3$	75
NaH_2PO_4	5
$NaSiO_3$	15

Vitamins	
Vitamin B_{12}	5×10^{-7}
Biotin	5×10^{-7}
Thiamin. HCl	1×10^{-4}

Trace metals	μM (final concentration)
Zn	0.08
Mn	0.9
Mo	0.03
Co	0.05
Cu	0.04
Fe	11.7
EDTA	11.7

Zn, Mn, Co and Cu may be added as chlorides or sulphates, and Mo is added as Na_2MoO_4. Fe and EDTA is conveniently supplied as Fe-EDTA. After autoclaving the seawater together with the nutrients the media has to rest for a few days to allow CO_2 to redissolve and the pH to decrease. The vitamins have to be prepared as a stock solution and sterile filtered (0.2 μm). The trace metals should be autoclaved as a stock solution separately, because otherwise precipitation might result. After addition of vitamins and trace-metals to the media the pH has to be adjusted to 7.4 (note that the pH in seawater is 8.2).

Some differences between fresh water and

saline algae systems (cultures or natural populations):

	Fresh water	Salt water
Typical limiting nutrients in nature	C, P	N
Typical planktonic algae	Chlorophytes Blue-green algae	Diatoms Dinoflagellates (Blue-green algae)
Typical pH	7	8.2
Method of concentration	Centrifugation Filtration	Filtration

PART II

BIOCHEMICAL ASPECTS OF PHOTOSYNTHESIS

CARBON METABOLISM

by J. COOMBS

4.1. INTRODUCTION

CO$_2$ is assimilated reductively by the enzyme ribulose *bis* phosphate carboxylase, and metabolised through the well-known Calvin cycle (Fig. 4.1) to result in a net synthesis of carbohydrate plus regeneration of the substrate ribulose *bis* phosphate.

In the C$_4$ plants CO$_2$ is initially assimilated by the enzyme phospho-enol pyruvate carboxylase. As a result the pattern and kinetics of labelling differs in these groups of plants. This can be followed using techniques for the assimilation and analysis of products of ^{14}CO$_2$. In general even simple fixation systems give good results. Typic-

ally, plants may be enclosed in plastic bags, or algae illuminated in simple chambers. ^{14}CO$_2$ is added after a period of equilibration, and the plant material killed and extracted in 80% v/v ethanol in water in a boiling water bath. The alcoholic solution is concentrated in a stream of N$_2$ gas or under vacuum and products separated using two-dimensional techniques of either paper chromatography in phenol water, followed by butanol acetic acid water, or more rapidly by electrophoretic acid chromatographic methods on cellulose thin layer plates. The nature of the products are established on the basis of their position and response to spray reagents: molybdate for sugar phosphates, ninhydrin for amino acids, pH dyes for organic acids and P-anisidine for carbohydrates.

If we consider the sequence of A → B → C and plot the percentage distribution of radioactivity in a given compound as a fraction of the total recovered against time for short periods (seconds) results will be as shown in Fig. 4.2.

For a C$_3$ plant, or most algae, A will correspond to phosphoglyceric acid, B to sugar phosphates and C to amino acids, organic acids and sugars. In

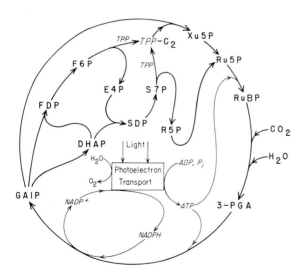

FIG. 4.1. The Calvin cycle. This may also be referred to as the photosynthetic carbon reduction (PCR) cycle or the reductive pentose phosphate pathway (RPPP).

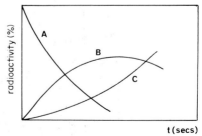

FIG. 4.2. Distribution of radioactivity in a sequence of compounds A → B → C with time.

contrast with C_4 plants the sequence of labelling will be found to be organic acids (or amino acid) to sugar phosphate to sugars. In C_4 plant extracts the first formed product (oxalo-acetic acid) may not be detected using a hot alcohol extraction as it is unstable. However, this compound may be trapped as the dinitrophenyl hydrazone.

4.2. SEPARATION PROCEDURES

Traditionally the products are separated using two-dimensional paper chromatography, and detected by autoradiography using X-ray film.

Ten to 100 μl are spotted on Whatman No. 4 chromatography paper (or equivalent) and run

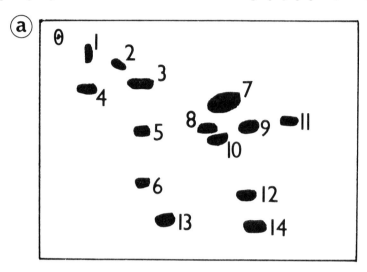

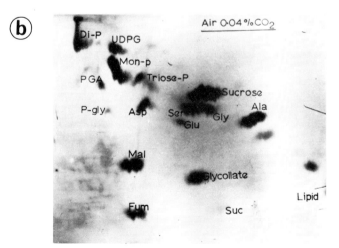

FIG. 4.3. Separation of products of $^{14}CO_2$ fixation, by photosynthesis in *Chlorella*, using two-dimensional paper chromatography in phenol:water followed by butanol:acetic acid:water. (a) Map showing expected position of major metabolites; (b) photograph of an autoradiograph obtained by placing a sheet of X-ray film in contact with such a paper chromatogram. Key: 1 = Diphosphate; 2 = Hexose monophosphate; 3 = Triose phosphate; 4 = Phosphoglyceric acid; 5 = Aspartate; 6 = Malate; 7 = Sucrose; 8 = Serine; 9 = Glycine; 10 = Glutamate; 11 = Alanine; 12 = Glycollate; 13 = Fumarate; 14 = Succinate.

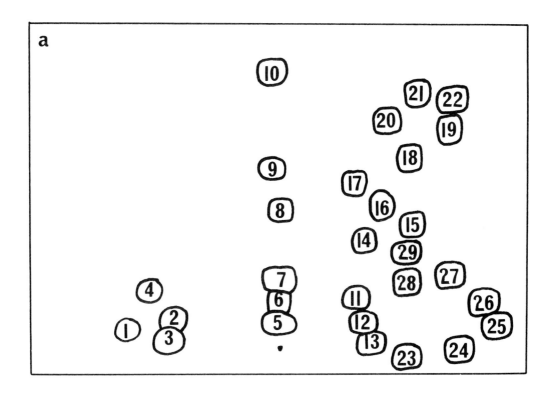

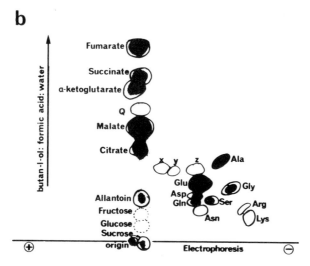

FIG. 4.4. (a) Map showing position of various metabolites following separation of plant extracts by 2-D. electrophoresis: TLC. *Key*: (1) Diphosphate, (2) Fructose 6.P, (3) Glucose 6.P, (4) PGA, (5) Sucrose, (6) Glucose, (7) Fructose, (8) Citrate, (9) Malate, (10) Fumarate, (11) Aspartate, (12) Glutamine, (13) Asparagine, (14) Glutamate, (15) Alanine, (16) Proline, (17) Tyrosine, (18) Methionine, (19) Valine, (20) Phenylalanine, (21) Leucine, (22) Isoleucine, (23) Cysteine, (24) Histidine, (25) Lysine, (26) Arginine, (27) Glycine, (28) Serine, (29) Threonine. (b) Autoradiograph of products of $^{14}CO_2$ fixation by detached nodules from *Phaseolus vulgaris*.

first in phenol: water (100 g phenol: 40 ml water), then in *n*-butanol: acetic acid: water (75:20:50). The separation is as shown in Fig. 4.3.

A much quicker procedure is to use a two-dimensional separation using low-voltage electrophoresis in the first direction and chromatography in the second. One to 10 μl of sample are spotted onto a pre-wetted cellulose thin-layer plate, and run for about 1.25 hours in 6 N formic acid. After drying it is then developed in the second dimension using butanol: formic acid: water (70:12:20). The distribution of products is as shown in Fig. 4.4.

4.3. DETECTION SPRAYS

Sugars. P-anisidine. Dissolve 0.5 g *p*-anisidine in 2 ml phosphoric acid, make up to 50 ml with ethanol, filter. After spraying, heat at 100°C for 5 minutes.

Organic acids. Bromocresol green. Dissolve 0.04 g bromocrestol green in 100 ml ethanol.

Add 0.1 M NaOH till blue colour just appears. Spray gives orange spots on blue background. (Plate must be heated at 120°C for 1 hour to remove residues of formic acid.)

Sugar phosphates. Molybdate perchloric acid. Dissolve 0.5 g ammonium molybdate in 5 ml H_2O. Add 1.5 ml 25% HCl and 2.5 ml 70% perchloric acid. Cool to room temperature and make up to 50 ml with acetone. Leave a day before use. Will keep 3 weeks. Develop spots with ultraviolet light.

Amino acids. Ninhydrin. Dissolve 0.5% w/v ninhydrin in *n*-butanol. Acidify with acetic acid, spray, heat.

4.4. PHOTORESPIRATION

In addition to catalysing the carboxylation of ribulose-bis-phosphate the RuBP carboxylase will also catalyse the oxygen-dependent formation of phosphoglycollate (oxygenase activity). This two-carbon compound is then metabolized through the C_2 pathway in the process known as

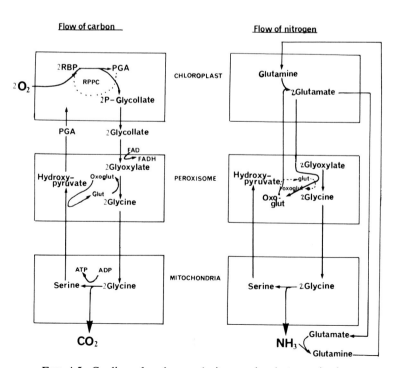

FIG. 4.5. Cycling of carbon and nitrogen in photorespiration.

photorespiration. During photorespiration the flow of carbon and nitrogen are as shown in Fig. 4.5. This figure emphasizes the importance of cycling of ammonia in photorespiration (Keys *et al.*, 1978) – a fact which is often overlooked. Since two molecules of glycine are used in the formation of one serine an imbalance results in the mitochondria. Similarly, two molecules of glyoxylate are aminated in the peroxisome, but only one amino group is available from the serine–hydroxypyruvate reaction. Hence the amino group released in the mitochondria must be recycled to the peroxisome. The ammonia released during the synthesis of serine is rapidly assimilated by cytoplasmic GS. However, the enzyme which catalyses the regeneration of glutamate as amino donor for glycine synthesis (GOGAT) is mainly present in the chloroplast. Hence, nitrogen must be recycled through the chloroplast. As a result of this activity the rate of net assimilation in C_3 plants is considerably reduced. However, due to technical difficulties the exact extent is not known. The problems arise from the fact that photorespiration represents the reverse process of photosynthesis, i.e.

photosynthesis	$CO_2 + H_2O \rightarrow CH_2O + O_2$
photorespiration	$CH_2O + O_2 \rightarrow CO_2 + H_2O$

Various factors affecting it may be summarized as shown in Fig. 4.6.

Light and Respiration

The rate of dark respiration may also be affected by the previous history of illumination. Respiration usually increases following periods of increasing illumination.

It is also possible that dark respiration is inhibited in the light.

Oxygen Uptake in the Light (Measurement by O^{18})

The uptake of O_2 in the light may be determined using the heavy isotope of oxygen, $^{18}O_2$. Experiments of this type have been used to suggest that respiration does not change in the light. However, for technical reasons such experiments are usually carried out under conditions of low O_2 (2%) and high concentrations of CO_2. As discussed later, these conditions do not favour photorespiration.

One has also to consider the extent to which the fed isotope is diluted with oxygen released in photosynthesis from the splitting of water. In other experiments the tissue has been illuminated under low light intensities, or algal cells bubbled with N_2. However, it has been shown that on increasing illumination $^{18}O_2$ uptake may

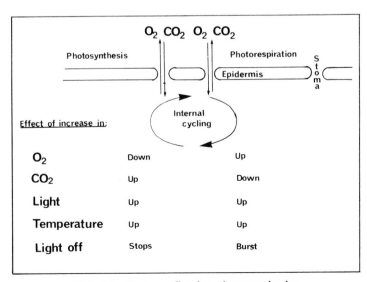

FIG. 4.6. Factors affecting photorespiration.

first decrease and then increase as light intensity is raised. The ^{18}O uptake was inhibited by the photosynthetic inhibitor DCMU.

To get meaningful results internal recycling must be minimal; furthermore, isotope experiments underestimate oxygen consumption. On the other hand, oxygen-consuming reactions under some conditions may not involve CO_2 production. This will lead to an overestimate of photorespiration.

Oxygen Inhibition

The rate of net photosynthesis can be increased in many (C_3) plants by lowering the oxygen concentration from 20% to 1%. It can also be decreased by raising the concentration to 100%. Several explanations are possible. This may be due to inhibition *per se*, or due to an increase in the loss of CO_2 due to photorespiration. An estimate of the magnitude of photorespiration can be made from data of this type.

Post-illumination CO_2 Burst

When an illuminated leaf is transferred to the dark an initial high rate of CO_2 evolution is observed. This may last for about 10 minutes. During this time rate of CO_2 evolution may be up to 4 times that of dark respiration. The magnitude of the CO_2 burst increases with light intensity, increases greatly with temperature and is not varied much with levels of CO_2. This overshoot would be expected if the substrate for photorespiration was formed in the light.

Extrapolation of Net Photosynthesis to Zero CO_2

It has been assumed that an extrapolation of rate of photosynthesis as a function of CO_2 concentration should indicate dark respiration if there was no dark fixation and if CO_2 production in light was equal to CO_2 production in dark.

Dilution of $^{14}CO_2$ Specific Activity in Ambient Atmosphere

If photosynthesis occurs in a closed system in the presence of $^{14}CO_2$, the CO_2 arising from

photorespiration will dilute the specific activity of the CO_2. This suggests photorespiration was 2 to 3 times dark respiration.

CO_2 (or $^{14}CO_2$ Loss Into CO_2-free Air or Other Atmospheres)

It has been suggested that at zero CO_2 concentration the efflux of CO_2 is a function of photorespiration, the diffusive resistance to CO_2 fixation in the mesophyll and the stomatal resistance. This technique is said to measure only a fraction of photorespiration and depends on (a) stomata wide open; (b) CO_2 concentration low; (c) rapid flow rate of CO_2. Apparent rate of photorespiration increases as gas-flow rate increases. This method has been used to confirm the importance of high light intensities and high O_2 partial pressures on photorespiration. Temperature effects are made complex since stomata begin to close at temperatures greater than 30°. The resulting increase in stomatal diffusive resistance leads to an apparent decrease in rate of photorespiration.

This method is not very sensitive, or accurate. However, if the tissue is first labelled with $^{14}CO_2$ the loss of radioactivity into the gas stream can be measured during subsequent periods of light or dark using a gas flow of the same rate.

In more detail the following simple methods give some indication of photorespiratory activity and are possible to relate to field studies whereas most of the above methods require complex instrumentation.

4.5. DETERMINATION OF THE "CO_2 COMPENSATION POINT"

4.5.1. Indicator Method (Qualitative)

Introduction

In a closed atmosphere, plants in the light will reduce the CO_2 concentration, of that atmosphere, until the CO_2 compensation concentration or point (Γ) is reached. The CO_2 compensation concentration (Γ) is normally expressed in cubic millimetres of CO_2 per litre of air ($mm^3 l^{-1}$). For C_3 plants $\Gamma = 50$ $mm^3 l^{-1}$ for C_4 plants $\Gamma = 0–10$ $mm^3 l^{-1}$.

In daylight green leaves have an influx of CO_2 due to photosynthesis greater than the efflux due to respiration. Consequently there is a net loss of CO_2 from the atmosphere to the leaves during the hours of daylight. Under field conditions, however, this change in CO_2 concentration is negligible as the perpetual turbulence of the atmosphere ensures that air depleted of CO_2 by the leaf is continually replaced by undepleted air. In a sealed vessel this replacement cannot occur and so the CO_2 concentration in the vessel is decreased. However, the rate of photosynthesis is partially dependent on CO_2 concentration. As the CO_2 concentration in a sealed vessel decreases, so the photosynthetic rate of the illuminated leaves will decrease until a point is reached where the rate of photosynthetic CO_2 influx to the leaf will equal the rate of respiratory CO_2 efflux from the leaf. Once this point is reached the CO_2 concentration of the vessel will remain constant. This is Γ, the CO_2 compensation point. Note that this respiratory CO_2 efflux is largely, if not entirely, due to photorespiration and not "dark" respiration which is negligible, or possibly inoperative, in illuminated green leaves. C_4 plants do not evolve CO_2 as a result of photorespiration, consequently they have a lower Γ than C_3 plants.

Method

A 250-cm^3 conical flask and bung provides a suitable sealed chamber. A good seal is absolutely essential. To each 250-cm^3 conical flask add 20 cm^3 of 5×10^{-3} M $KHCO_3$ and 5 drops of Universal Indicator, as shown in Fig. 4.7.

Cut the ends of the leaves or shoot supplied underwater, and transfer the cut ends to small vials underwater. Leave sufficient water in the vials for the ends to be immersed. Tie a thread around each vial and lower each into a flask. Seal each flask with a bung. An additional seal with Propafilm, if available, would be advantageous. Place each flask within about 30 cm of one or two 60 W tungsten lamps (or under a bank of neon tubes, but again within 30 cm). Leave the flask for at least 1 hour. During this

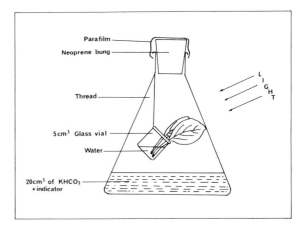

FIG. 4.7. Experimental setup for qualitative determination of the CO_2 compensation point using the indicator method.

time examine the anatomy of the leaves of the plants that you are investigating. After $1\frac{1}{2}$–2 hours the pH of the solution in the flask is read by comparison of its colour with that of the Universal Indicator colour chart.

This method is really only qualitative. However, in theory some indication of the compensation point can be calculated as follows:

Calculation of Results

From the pH of the bicarbonate solution the CO_2 concentration in the flask, and hence the CO_2 compensation point, may be determined.

Dissolved CO_2 or carbonic acid dissociates to form hydrogen ions and bicarbonate ions.

$$H_2CO_3 \rightleftharpoons H^+ + HCO_3^-$$

By definition the dissociation constant (K) for dissociation is

$$K_1 = \frac{[H^+][HCO_3^-]}{H_2CO_3} \text{ in moles}$$

As less than 1% of dissolved CO_2 in fact exists as carbonic acid, we may write:

$$K_2 = \frac{[H^+][HCO_3^-]}{CO_2}$$

$$\therefore [H^+] = \frac{K_2[CO_2]}{[HCO_3^-]} \qquad (1)$$

Table 4.1. *Solubility of CO_2 in water (α)*

Temp. (°C)	Solubility (m^3 CO_2 m^{-3} of water)
15	0.988
20	0.848
25	0.730
30	0.652
35	0.570

By definition $pH = -\log_{10}[H^+]$

Similarly $pK = -\log_{10}[K]$

Taking log to the base 10 in (1) we obtain

$$pH = pK_2 - \log_{10}\frac{[CO_2]}{[HCO_3^-]}$$

$$\therefore pH = pK_2 - (\log_{10}[CO_2] - \log_{10}[HCO_3^-])$$

$$\therefore \log_{10}[CO_2] = pK_2 + \log_{10}[HCO_3^-] - pH$$

From this equation the molar concentration of CO_2 can be calculated, given that $pK_2 = 6.37$ for the dissociation of carbonic acid and assuming that the decrease in HCO_3^- concentration is negligible. The concentration of CO_2 in the atmosphere of the flask may then be obtained from the equation below

$$C_a = \frac{[CO_2] \times 22.4}{\alpha} \times 10^6$$

where [CO_2] is the molar concentration of CO_2,
    22.4 is the gas-mole conversion factor,
    α is the solubility of CO_2 for a given temperature, in m^3 CO_2 m^{-3} water (Table 4.1),
    C_a is the concentration of CO_2 in air (cm^3 m^{-3}).

4.5.2. Mylarbag Assay* (Semi-Quantitative)

Mylar is the registered trademark for a polyester film made by E.I. du Pont de

*Goldsworthy (1970).

Nemours & Co., Inc. Mylar bags used for this work are obtained from Vac Pac, Inc., Box 6339, Baltimore, Maryland 21203.

Two replicate portions of leaf material, each of approximately $60 \, cm^2$ (the exact size is not critical), are used for each determination. They may take the form of 8×8 cm squares of leaf tissue cut from either side of the midrib of the same leaf, but whole detached leaves of other plants may be used with equal effect. The leaf material for each replicate is floated, undersurface uppermost, on 15 ml of distilled water at the desired temperature in an open 10-cm-square plastic Petri dish. Each dish is slid horizontally to the far end of a 16×45-cm Mylar bag. One of the replicate bags is pumped up with atmospheric air and closed half-way along its length by twisting tightly and sealing the twist with a burette pinchcock. The remaining bag is flushed by filling it three times with CO_2-free air and expelling the contents. It is then filled with CO_2-free air, twisted half-way down, and sealed with a pinchcock. Both bags are laid horizontally in a growth chamber under defined conditions of temperature and illumination for a period of 1 hour. The gas atmosphere in the bag is then sampled while still under illumination and analysed for CO_2 concentration.

The technique for sampling the gas atmosphere in the bags is as follows: A sampling probe made from a 5-ml graduated pipette is connected via a small magnesium perchlorate drying column to an infra-red CO_2 analyser. The volume of the whole system, including the analyser, should be less than 200 ml. The tip of the pipette is inserted into the mouth of the bag down as far as the pinchcock. The neck of the bag is then twisted tightly around the body of the pipette *in the opposite direction to the twist originally employed to seal the bag*. The neck of the bag is gripped firmly, the pinchcock removed and the body of the bag squeezed. The complementary nature of the two twists allows the bag to open in the position vacated by the pinchcock and the gas atmosphere within is expelled by manual pressure into the CO_2 analyser.

4.6. LIGHT/DARK ^{14}C ASSAY FOR PHOTORESPIRATION*

Leaves weighing about 10 g are excised and placed with their bases in water in the dark for 1 to 4 hours before leaf disks 1.6 cm in diameter are cut from symmetrical positions with a sharp punch. The disks are floated on water and six disks (about 240 mg fr. wt.) strung by means of a needle on thread to facilitate rapid handling. The sets of disks are placed in large (75 ml) Warburg vessels containing sufficient water to just cover the bottom of the vessel (1.5 ml). Usually two vessels are compared at the same time in each experiment. The vessels are attached to Warburg manometers and shaken for 45 minutes at 120 oscillations per minute in a constant-temperature bath at 30° or 35° in the light (1000–200 ft-c supplied by tungsten lamps above and a mirror directly below the vessels) with the vessels open to the air. During this preliminary period, the stomata open and subsequent $^{14}CO_2$ uptake occurs at a rapid rate.

The vessels are then closed to the atmosphere and 5 μmoles of $^{14}CO_2$ of known radioactivity are liberated by the injection of acid through a rubber serum stopper in the sidearm to $NaH^{14}CO_3$ solution. The $^{14}CO_2$ is entirely assimilated by the leaf disks in about 15 minutes. However, a total of 45 minutes is allowed to elapse in the closed system before the start of the photorespiration measurements at zero time.

Photorespiration Assay

A venting plug is inserted into the sidearm of the Warburg vessel, and tubing connected to the top of the manometer. Air is passed in succession through an aeration tube, through 250 ml of 2.5 N KOH, then through water and finally circulated through coiled copper tubing placed in the constant temperature bath. The CO_2-free air at the bath temperature is then led into the vessel through the top of the manometer and out through the venting plug into an aeration tube inserted in 50 ml of 1.0 M ethanolamine to trap the $^{14}CO_2$ released. The flow rate of the CO_2-free air passing over the leaf disks is adjusted to 500 ml per minute (6.7 flask volumes per min) and 0.25-ml samples removed periodically from the ethanolamine solution. The radioactivity of these is determined by scintillation counting.

After 35 minutes, the lights are turned off and the $^{14}CO_2$ released in the dark (less than 10 ft-c) determined. To calculate the ratio of photorespiration to dark respiration, the radioactivity in the $^{14}CO_2$ released in the period between 5 and 35 minutes in the light and between 15 and 45 minutes in the dark is generally used.

Comments on Compensation Point and Photorespiration Assays

The above methods are attempts to obtain results of a very complex phenomenon using simple apparatus—hence all are subject to some doubt if used for detailed comparative comparison of plants of different species or types. For instance, in the flask method the environment is "unreal". For the bag experiment temperature control must be precise if the method is to yield quantitative results. A time course experiment should be run for each species and each light intensity to establish that a steady state has been reached. As far as the ^{14}C assay is concerned there have been attempts to interpret the ratio of ^{14}C evolved in the light to that evolved in the dark in a quantitative way—i.e. to indicate different activities of photorespiration in different species. There is no basis for such comparison.

Bibliography and Further Reading

BASSHAM, J. A. and CALVIN, M. (1957) *The Path of Carbon in Photosynthesis.* (Prentice–Hall: Englewood Cliffs, N.J.)

BURRIS, R. H. and BLACK, C. C., eds. (1976) *CO_2 Metabolism and Plant Productivity.* (University Park Press: Baltimore.)

GIBBS, M. and LATZKO, E., eds. (1979) Photosyn-

*Zelitch, I., *Plant Physiol.* **43**, 1829–1837 (1968).

thetic carbon metabolism and related processes. Photosynthesis II. *Encyclopedia of Plant Physiology*, new series, vol. 6. (Springer-Verlag.)

GOLDSWORTHY, A. and DAY, P. R. (1970) A simple technique for the rapid determination of plant CO_2 compensation points. *Plant Physiol.* **46**, 1829.

HATCH, M. D. and SLACK, C. R. (1970) Photosynthetic CO_2 fixation pathways. *Ann. Rev. Plant Physiol.* **21**, 141.

HATCH, M. D., OSMOND, C. B. and SLATYER, R. O., eds. (1971) *Photosynthesis and Photorespiration.* (Wiley Interscience: N.Y., London.)

KEYS, A. J., BIRD, I. F., CORNELIUS, M. WALLSGROVE, R. M., LEA, P. J. and MIFLIN, B. J. (1978) Photorespiratory nitrogen cycle. *Nature* **275**, 741.

SIEGELMAN, H. W. and HIND, G., eds. (1978) *Photosynthetic Carbon Assimilation.* (Plenum Press: N.Y., London.)

TOLBERT, N. E. (1971) Microbodies, peroxisomes and glyoxisomes. *Ann. Rev. Plant Physiol.* **22**, 45.

ZELITCH, I. (1968) Investigations on photorespiration with a sensitive ^{14}C-assay. *Plant Physiol.* **43**, 1829.

ZELITCH, I. (1971) *Photosynthesis, Photorespiration and Plant Productivity.* (Academic Press: N.Y.)

AN EXPERIMENT TO TEST QUANTITATIVE TECHNIQUES

by G. HIND

Experimental Procedure

Cytochrome c is a haemoprotein which functions as an electron carrier in respiration. Its oxidized and reduced forms have markedly different absorption spectra in the visible wavelengths. Spectrophotometry can thus be used to determine both the amount of cytochrome present and the percentage that is in the reduced state. The spectrophotometric technique will be used here to test the accuracy of solutions made up by weight and diluted.

Given that the molecular weight of this cytochrome c is 12,400, prepare 10 ml of a 20 micromolar solution by use of the volumetric flask and analytical balance provided. By use of appropriate pipettes, dilute this stock solution to give 5 ml each of a 10, 5, 2 and 1 micromolar solution. Set the spectrophotometer wavelength to 550 nm and measure the absorbance of the stock solution relative to a reference cell filled with water. If the instrument has a scanning drive, scan over the range 400 to 600 nm and record the absorption spectrum of oxidized cytochrome c. Now add to the same sample a few grains of sodium dithionite, cover the cuvette with plastic film, mix by inverting, then measure the absorbance again at 550 nm and if possible record the spectrum of the reduced cytochrome on the same sheet of paper. Repeat this procedure, taking readings at 550 nm only, with the diluted samples.

For a cuvette presenting a 1 cm depth of sample to the spectrophotometer measuring beam, the absorbance (A) is related to the concentration (c) of absorbing molecules by: $c = A/e$, where e is the extinction coefficient at the wavelength of measurement. If the millimolar coefficient is used, the units of concentration will be millimolar (mM). Given that:

for oxidized cytochrome c at 550 nm, $e = 10.5 \ (\text{mM} \times \text{cm})^{-1}$,

for reduced cytochrome c at 550 nm, $e = 28.0 \ (\text{mM} \times \text{cm})^{-1}$,

calculate the theoretical absorbance of the stock solution before and after reduction and use these values to illustrate on graph paper the theoretical relationship between absorbance at 550 nm (vertical axis) and concentration of the fully reduced and fully oxidized cytochrome. On this graph show also, as separate points, the absorbances actually measured for the stock solution and its dilutions.

What was your worst percentage deviation from the theoretical value? Were the errors randomly scattered around the theoretical line (pipetting error), on a different straight line (error in preparing the stock solution) or scattered around a different line (both errors)?

The difference between the extinction coefficients for two states (e.g. oxidized and reduced) of the same substance is called the difference extinction coefficient. Here the value of $17.5 \ (\text{mM} \times \text{cm})^{-1}$ is the reduced minus oxidized difference coefficient at 550 nm for cytochrome c. The solution provided is a mixture of oxidized cytochrome c and chloroplast pigments. These absorb throughout the visible region, so the concentration of cytochrome c cannot be found by simple use of the extinction coefficient. Provided, however, there are no other substances present that change absorbance at 550 nm upon adding dithionite, the difference coefficient provides a means of selectively measuring the cytochrome c content. Attempt to measure the cytochrome c content of the mixed solution using this information.

Chloroplasts contain cytochrome f, which resembles cytochrome c. It is closely associated with photosystem I which oxidizes it, and it is reduced with electrons from photosystem II. With the aid of specially sensitive spectrophotometers, its functioning during photosynthesis can be observed and its concentration can be determined by use of the difference extinction coefficient as in the above exercise.

APPENDIX A.5. SPECTROSCOPY

The Absorption of Light by Molecules

A beam of light passing through a substance can be absorbed if the energy of the photons happens to coincide with a particular energy transition available to the molecules. So if monochromatic light of varying wavelength is passed through a substance, the absorption of that light will rise to a peak value and then fall again as the wavelength coinciding with an energy transition is reached and passed. In some substances several such peaks may be found and may superimpose to form a broad peak. If the amount of absorption is measured and plotted against wavelength the resulting curve is called an absorption spectrum and is characteristic of the absorbing substance. The absorption at a wavelength where absorption is greatest is termed the peak extinction or extinction maximum.

If absorption is measured at a fixed wavelength, the value obtained will depend on three factors. Firstly, the thickness of absorbing matter through which the light passes; secondly, the concentration of absorbing matter present; and thirdly, the efficiency with which the matter can absorb light.

The Measurement of Light Absorption

A conventional laboratory spectrophotometer measures the intensity of monochromatic light which has passed through a sample, with the objective of determining how much light was absorbed by molecules in solution or dispersion.

Since absorption varies with wavelength it is essential to have a light source and monochromator which together produce a beam of light with a narrow band of frequencies. The intensity of light in this beam will vary at different wavelengths because the spectral distribution of energy from the lamp is uneven. Also the efficiency of the photodetector is wavelength dependent. In addition, since the sample is a solution which must be contained in a suitable vessel (the cell), a reference measurement is necessary to compensate for absorption by the solvent and cell.

In practice two identical cells are used: one, the reference cell, containing only the solvent, the other, the sample cell, containing the test solution. At a given wavelength the photodetector will respond linearly to the amount of light falling on it, so the reference cell and then the sample cell are placed into the beam one after another and the ratio of the amount of light transmitted by the two of them is determined. This ratio is called the fractional transmission of the sample (or when multiplied by 100, the percentage transmission). In many modern spectrophotometers the measuring beam is split and chopped so that light is presented alternately to the sample and reference cuvettes without action on the part of the operator.

For a thin layer, the same proportion of the incident light is absorbed whatever the incident intensity (Lambert's Law).

A sample cell can be thought of as being divided into a number of thin layers, each successive layer absorbing the same proportion of the light falling on it. Thus if the first layer absorbs half the light, the second layer will absorb half the remaining light, i.e. a quarter of the total initial light. Extrapolating this argument to layers of infinite thinness the light intensity will decrease exponentially through the cell. This may be expressed as:

$$\ln \frac{I_0}{I} = a \cdot l$$

where I_0 is the incident light falling on the cell, I is the emergent light, a is a constant, l is the path length of the cell.

Beer's Law states that light absorption is proportional to the number of absorbing molecules. Using an argument similar to that above:

$$\ln \frac{I_0}{I} \propto c$$

where c is the concentration of the solute molecules.

Combining these two laws we obtain:

$$\ln \frac{I_0}{I} = a.c.l. \text{ or } \log_{10} \frac{I_0}{I} = \frac{a.c.l.}{2.303} = 0.434 \ a.c.l.$$

If c is the molar concentration of the solute and l is in centimetres then $0.43a$ is a constant called the molar extinction coefficient normally given the symbol e.

$\log_{10} I_0/I$ is termed the absorbance (A) of the solution and is directly proportional to the concentration of solute if the path length is kept constant. Thus the expression above becomes:

$$A = e.c.l.$$

Using this expression it is possible to calculate the concentration of a solution, if the extinction coefficient of the solute and the path length of the cell are known, simply by measuring the absorbance. The extinction coefficient is given by:

$$e = \frac{A}{c.l}.$$

A is a ratio and is therefore dimensionless, c has the dimensions of molar (gram molecules/litre) and l has dimensions of cm, therefore e has dimensions of $(\text{M} \times \text{cm})^{-1}$. Since e varies with wavelength it is normally quoted at the peak extinction wavelength.

The ratio $(\log_{10} I_0/I)$ is normally called absorbance (A); other terms are in common use, namely "optical density" OD and "extinction" E. These may be numerically equivalent but absorbance is now preferred. Another term in use is percent transmittance $(\%T)$ which is equal to $100 . I/I_0$. Changes in transmittance or absorbance are referred to as ΔT and ΔA respectively. If these are very small ($< 1\%$) then $\Delta A \approx 0.434\Delta T$. The effect of this is important in the design of sensitive difference spectrophotometers because the difference in the output current of the photodetector is linearly related to ΔT and thus to ΔA—if these are small; hence complex logarithm-taking circuitry is unnecessary.

Section 6

CHLOROPLASTS AND PROTOPLASTS

by R. C. LEEGOOD, G. E. EDWARDS and D. A. WALKER

6.1. INTRODUCTION

A large part of our present knowledge concerning the nature and regulation of processes such as electron transport, photophosphorylation and CO_2 assimilation derives from experiments with isolated chloroplasts. It is widely accepted that fully functional organelles provide a more realistic basis for studies of such processes than those which have been unintentionally stripped of their limiting envelopes during isolation. Chloroplasts which are largely intact and display an ability to assimilate carbon dioxide at rates which match those of the parent tissue are regarded as metabolically competent. Although lack of damage to chloroplasts during isolation procedures cannot be guaranteed at present, envelope integrity is easily assessed since it bears an inverse relation to the ability of isolated plastids to support ferricyanide-dependent O_2 evolution. Until relatively recently, intact chloroplasts which fix CO_2 at high rates have only been separated routinely from spinach and peas by mechanical disruption of the tissue. Other species have usually been found unsuitable for this purpose because of the presence of phenolics in their leaves or because they possess a high proportion of tissue with thickened cell walls.

The introduction and use of protoplasts, though still in its infancy, has greatly extended the range of plants from which intact chloroplasts may be isolated. The technique has been used successfully with C_3 grasses such as wheat and barley and with some C_3 dicotyledons such as sunflower, tobacco, peas and spinach, as well as many C_4 species. In certain respects this method

of preparing chloroplasts may be considered superior to mechanical disruption. The mechanical force required to rupture the plasma membrane is much less than that required to break cell walls, so that disruption greatly reduces damage to organelles. Chloroplasts isolated from protoplasts thus show very high intactness. Secondly, interfering substances present in the cell walls or vascular tissues are removed during preparation of the protoplasts. These would be released during the mechanical maceration of the tissue, and their effects can only be minimized by rapid centrifugation of the organelles. For example, with sunflower, a species with phenolics present in the leaves, mechanical procedures yield chloroplasts which assimilate CO_2 at rather low rates despite the inclusion of a great many protective agents in the isolation medium. Protoplasts, in contrast, yield chloroplasts of high activity. Similarly, it is impossible to prepare chloroplasts by mechanical means from flag leaves of wheat, whereas good chloroplasts can be successfully prepared from protoplasts.

The isolation of organelles with a high degree of integrity is of particular value in studies of the intracellular location of enzymes. Thus protoplasts have been a valuable tool in the elucidation of the relative roles of mesophyll and bundle sheath cells in C_4 plants. More recently, a method has been developed for the isolation of chloroplasts from illuminated protoplasts (Robinson and Walker, 1979), allowing studies of the intracellular distribution of metabolites during photosynthesis. Intact protoplasts also permit the study of photosynthesis in a system which avoids the limitations of the intact leaf (e.g.

stomatal limitation of gas exchange), although it should be borne in mind that rates of transport between the cell and the medium are probably much reduced (certainly in C_3 species) compared with rates of transport between cells in the intact tissue. There is also the possibility that some features of metabolism may be the result of a wound response.

Whichever method of preparation is chosen, it is impossible to isolate protoplasts or chloroplasts with high rates of photosynthesis from poor or indifferent material. Even spinach and peas will only yield "good" chloroplasts if the parent tissue is itself in "good" condition. Thus adequately controlled growth facilities are essential to allow reasonably reproducible results throughout the year. Spinach is a particularly demanding plant in this respect, requiring a short daylength, together with a high light intensity, high humidity and a moderate temperature. Good results have been obtained with plants grown in water culture. As plants are grown for 6–8 weeks, a reasonably large stock is needed. Peas, usually of a dwarf variety, can be grown more simply in a medium like vermiculite, until about 10 to 12 days old. They are best grown under low light intensities in order to reduce starch formation. Whole young shoots may be employed in mechanical isolation procedures and young expanded leaves for protoplast preparations. Wheat and barley grown, like peas, for 6 to 10 days in vermiculite watered with a nutrient solution are ideal material for protoplast isolation. Slightly higher light intensities are used than for peas (about 30 W m^{-2} in addition to natural light). As for spinach, once the best system, including the variety, has been established, this is best adhered to. Small variations in growth conditions may bring about quite marked changes, for example, in susceptibility to enzymic digestion or in rates of photosynthesis by isolated protoplasts and chloroplasts.

There can be little doubt that time and persistence will lead to an extension of the number of species from which protoplasts and chloroplasts may be isolated by modifications of existing methods, although it is not recommended that experience in either method of preparation be gained with anything other than established species.

6.2. ISOLATION OF PROTOPLASTS

Preparation of Leaf Tissue

The method used for preparing leaf tissue for digestion depends upon the species used. With monocots, such as wheat, barley and maize, leaf segments can be cut using a razor-blade (transverse segments of about 0.5–1.0 mm). Alternatively a mechanical leaf-cutter may be used (Huber and Edwards, 1975). With dicots, the epidermis can sometimes be removed from the lower surface of the leaf (e.g. tobacco, peas, spinach, *Kalanchoe daigremontiana*). The surface of the leaf can be rubbed with carborundum or brushed gently (e.g. with a tooth-brush) in order to break through the epidermal tissue, a method particularly successful with sunflower. In cases where air might enter the leaf during cutting, either vacuum infiltration or else cutting the tissue into thin segments under 0.5 M sorbitol (which works well with spinach) should be considered.

Once prepared, the leaf tissue is incubated for a maximum of about 4 hours at 25–30°C under low light. Longer incubations are undesirable, as they may lead to a loss of photosynthetic function. In certain cases, shaking during the incubation may improve digestion, but it does not seem to be of benefit when thin leaf sections of monocots are used.

The enzymes used in incubation are commercially available (see Materials, Section 6.9), and have been found particularly useful with a number of species including most monocots. For species resistant to digestion, either some variation in growth conditions or alternative sources of digestive enzymes may prove useful, e.g. Rohament P pectinase is particularly effective with sunflower leaves.

Isolation and Purification

Following incubation of the leaf tissue with digestive enzymes, the isolation medium can be gently removed and discarded. Successive washing with an osmoticum (e.g. in 0.5 M sorbitol, 1 mM $CaCl_2$) releases a mixture of protoplasts, chloroplasts and vascular tissue. Undigested material including undigested tissue, epidermal tissue and vascular strands can be removed by filtration through nylon sieves. Bundle-sheath strands from C_4 plants are resistant to digestion and can be collected on 80-μm nylon mesh. Low-speed centrifugation of the extract, at 100 g for 5 minutes, gives a pellet containing protoplasts and chloroplasts. There are two means of purifying protoplasts. With some species mesophyll protoplasts will float when resuspended in a solution of sufficiently high density. For example, if mesophyll protoplasts of wheat, barley or sunflower are suspended in a solution containing 0.5 M sucrose and centrifuged at low speeds (250 g for 5 minutes), they float to the top of the medium. Layering a lower density osmoticum (e.g. sorbitol) on top of the sucrose medium prior to centrifugation results in protoplasts partitioning at the interface. They are then easily collected with a Pasteur pipette (Fig. 6.1).

Protoplasts which do not float readily can be purified in two ways. Dextran T_{20} can be added to 0.5 M sucrose medium at a concentration of 5–10% (dextran is a polymer which increases the density of the medium), or the protoplasts can be purified in an aqueous two-phase system formed by a mixture of two polymers: dextran and polyethylene glycol. The actual mixture is composed of 5.5% (w/v) polyethylene glycol 6000, 10% (w/v) dextran T_{20}, 10 mM sodium phosphate (pH 7.5) and 0.46 M sorbitol. If 0.6 ml of the unpurified protoplast preparation is thoroughly mixed with 5.4 ml of the two-phase solution and centrifuged at 300 g for 5 minutes at 4°C, the protoplasts partition at the interphase and chloroplasts in the lower phase. Protoplasts collected from the interphase can be resuspended and washed in an appropriate storage medium (Kanai and Edwards, 1973).

6.3. MESOPHYLL PROTOPLASTS FROM C_3, C_4 AND CAM PLANTS

Make up the following digestion medium (suitable for young leaves of wheat (*Triticum aestivum*) and CAM species).

    2% (w/v) cellulase (Onozuka 3S),
    0.3% (w/v) pectinase (Macerozyme R-10),
    0.5 M sorbitol,
    1 mM $CaCl_2$,
    5 mM MES.
    Adjust to pH 5.5 with dilute HCl.

Usually digestion is done at pH 5.0–5.5, a compromise between the pH optimum for the enzymes (which is more acid) and avoidance of damage to the tissue. With some tissues, such as maize, it may be beneficial to use a more alkaline digestion medium (pH 6.0–7.0) in order to retain maximum photosynthetic activity, although the yield will, of course, be reduced. Variations in digestion media:

Spinach (*Spinacia oleracea*); 3% Onozuka 3S, 0.5% Macerozyme R-10.

Maize (*Zea mays*) and C_4 sp.; 2% Onozuka 3S, 0.1% Macerozyme R-10.

Sunflower (*Helianthus annus*); 2% Onozuka 3S, 0.5% Rohament P.

Pea (*Pisum sativum*); 2.5% Onozuka 3S, 0.5% Macerozyme R-10, 0.025% Rohament P.

0.05% bovine serum albumin (defatted) can also be used as a wetting agent.

Cut up leaf tissue with a sharp razor-blade into segments approximately 1 mm wide for CAM, 0.7 mm for C_3 and 0.5 mm for C_4. (Note: it is important to cut small segments of C_4 leaves in order for bundle sheath strands to be released during the digestion period.) Put about 3 g of tissue with 20 ml of enzyme medium in a 9 cm diameter petri dish. Incubate under low

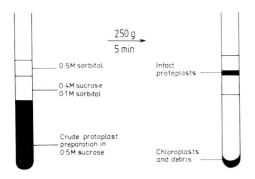

FIG. 6.1. Scheme for the purification of protoplasts by flotation in a step-gradient of sucrose and sorbitol.

light at 25–30°C for 3–4 hours. (For maximum photosynthetic activity of maize mesophyll protoplasts, a shorter incubation of about 2 hours may be necessary.) Gently remove the enzyme medium from the dish using a pipette. Several petri dishes of tissue should provide an adequate yield.

C₃ Tissue

Make up the following media and use them chilled throughout the protoplast purification:

I   0.5 M sorbitol
     1 mM $CaCl_2$
     5 mM MES, pH 6.0
II   0.4 M sucrose
     0.1 M sorbitol
     1.0 mM $CaCl_2$
     5 mM MES, pH 6.0
III 0.5 M sucrose
     1 mM $CaCl_2$
     5 mM MES, pH 6.0

Gently wash the segments in 20 ml medium I. Filter the extract through 500-μm nylon mesh (e.g. a tea-strainer) and 200-μm nylon mesh. Wash the segments twice more and centrifuge the combined extract in 15×20 mm tubes at 100 g for 5 minutes. Discard the supernatant and gently resuspend the pellet in a few drops of sucrose medium (III). Add 5 ml medium III then layer 2 ml medium II on top (this washes the protoplasts), followed by a layer of 1 ml of sorbitol medium (I). Centrifuge at 250 g for 5 minutes and collect the pure protoplast fraction at the interface with a Pasteur pipette. Usually the pellet consists of debris (chloroplasts and broken cells) but it is worth checking by light microscopy that no protoplasts are discarded. If many protoplasts are present in the pellet, some dextran T₂₀ (5–10% (w/v)) may be added to the sucrose medium and the purification repeated.

C₄ Tissue

The purification procedure is similar to that of C₃ protoplasts with the following modifications. After washing the segments, filter the extract through 500-μm and 80-μm nylon mesh. The

bundle-sheath strands will collect on the 80-μm mesh. Resuspend the bundle-sheath strands in sorbitol medium (I). The mesophyll protoplasts can be purified from the filtrate by a method similar to that used for C₃ protoplasts. The sucrose medium should contain dextran T₂₀ (10% (w/v)). (C₄ mesophyll protoplasts are generally more dense than those of C₃ leaves, thus a denser medium is required for flotation.) A layer of 1 ml sorbitol medium should be put directly on top of the sucrose layer.

CAM Tissue

Attempts to isolate cells should only include pectinase in the digestion medium. Attempts to isolate protoplasts should have pectinase and cellulase in the digestion medium.

For cells, wash segments after digestion with a medium containing 0.5 M sorbitol, 1 mM $CaCl_2$, 50 mM Tricine-KOH, pH 7.8 (sorbitol medium). Filter through 1-mm and 500-μm nylon mesh. Centrifuge cells at 100 g for 3 minutes and resuspend in the sorbitol medium followed by storage on ice.

In the case of protoplasts, overlay the extract onto a cushion of 0.5 M sucrose, 1 mM $CaCl_2$, 50 mM Tricine-KOH, pH 7.8 and 20% (w/v) dextran T₂₀. Alternatively, because protoplasts from CAM species are usually larger and more dense then those from C₃ and C₄ species, they can be purified by allowing the intact protoplasts to settle out. The supernatant may then be discarded.

Storage of Protoplasts

For storage, measurements of photosynthesis and for chloroplast preparation the protoplasts removed from the gradient should be diluted with sorbitol medium (about 10-fold) and centrifuged at 100 g for 5 minutes. Protoplasts for immediate use can be resuspended in sorbitol medium and stored on ice.

Mesophyll protoplasts are most stable if stored at a relatively low pH and with some divalent cations (e.g. 0.4 M sorbitol, 1 mM $CaCl_2$ and 20 mM MES, pH 6.0). (Studies with barley, wheat and tobacco indicate that sorbitol is an adequate osmoticum for at least 8 to 10 hours if the protoplasts are stored on ice. If stored at 25°C, the protoplasts lose their photosynthetic capacity when a sorbitol medium is used. However, the protoplasts are photosynthetically stable for at least 10 to 20 hours at 25°C with sucrose as the osmoticum.)

6.4. PHOTOSYNTHESIS BY ISOLATED PROTOPLASTS

C₃ Protoplasts

When using the oxygen electrode, care should be taken to set the stirrer at the lowest setting consistent with an adquate response of the electrode to the oxygen evolved. In this way breakage of the fragile protoplasts is minimized. Protoplasts can be illuminated with white light, a suitable concentration is 50 μg Chl/ml. A wide range of media may be used for the assay of CO_2-dependent O_2 evolution. The pH optimum is broad as long as the concentration of bicarbonate is kept low at acid pH values ($<$ pH 7.0). $CaCl_2$ is generally included not only to prevent clumping but also to inhibit photosynthesis by chloroplasts which are inevitably present in the preparation as a result of rupture of protoplasts. A suitable medium for wheat protoplasts is 0.4 M sorbitol, 10 mM $NaHCO_3$, 5 mM $CaCl_2$, 50 mM Tricine-KOH, pH 7.6 at 20°C. $^{14}CO_2$ fixation may be measured in the same medium.

C₄ Mesophyll Protoplasts

O_2 evolution: resuspend mesophyll protoplasts at a concentration of about 50 μg Chl/ml in 0.33 M sorbitol, 1 mM $MgCl_2$, 1 mM $MnCl_2$, 1 mM EDTA, 2 mM KH_2PO_4, 50 mM Tricine-KOH, pH 7.5. Conduct experiments at 35°C.

CO_2 fixation: use the same medium as for O_2 evolution experiments. Assays can be run in a volume of 250 μl. Do the assays at two chlorophyll concentrations (e.g. 5 and 10 μg Chl/assay). Use 6 mM $NaH^{14}CO_3$ in the assay (approximately 1 μCi). Take 5 μl aliquots at 5, 10, 15 and 25 minutes and add to scintillation vials containing 0.1 ml 20% (w/v) trichloroacetic acid (TCA) and 0.1 ml of 10 mM phenylhydrazine.

The TCA will acidify the extract and release any unfixed $^{14}CO_2$, while the phenylhydrazine will tend to stabilize any ^{14}C-labelled oxaloacetate by forming the phenylhydrazone derivative. Oxaloacetate is labile, decarboxylating to pyruvate and CO_2, particularly in the presence of divalent cations or cellular constituents. Allow the scintillation vials to stand for 30 minutes, flush with air and add 10 ml scintillation fluid (e.g. 70% v/v toluene, 30% (v/v) ethanol, 6 g/l PPO (2, 5-diphenyloxazole), 300 mg/l POPOP (1, 4-bis(2(5-phenyloxazolyl)) benzene).

Plot dpm fixed versus time for each experiment. In the linear phase of CO_2 fixation, determine dpm fixed for each 5-minute interval. Then, from the specific activity of the $^{14}CO_2$, calculate the rate of CO_2 fixation as μmol CO_2 fixed mg^{-1} Chl h^{-1}.

The following treatments should be considered, and experiments run in the light and dark. For dark treatments, wrap the tubes in aluminium foil. Pyruvate and oxaloacetate should be dissolved just prior to use and adjusted to pH 7–8.

1. No substrates.
2. 5 mM pyruvate.
3. 5 mM ribose 5-phosphate.
4. 5 mM phosphoenolpyruvate.
5. 5 mM pyruvate, 1 mM oxaloacetate.
6. 5 mM 3-phosphoglycerate.
7. 5 mM pyruvate, 1 mM oxaloacetate, 10 mM D, L-glyceraldehyde.
8. 5 mM pyruvate, 1 mM oxaloacetate, 5 mM malonate.

C₄ Bundle-sheath Strands

Light-dependent O_2 evolution may be measured with the addition of various substrates in a medium containing 0.33 M sorbitol, 1 mM $MgCl_2$, 1 mM $MnCl_2$, 1 mM EDTA, 2 mM KH_2PO_4, 50 mM Tricine-KOH, pH 7.5. Assays that might be included:

1. 10 mM $NaHCO_3$.
2. 10 mM $NaHCO_3$, 10 mM ribose 5-phosphate.
3. 10 mM $NaHCO_3$, 25 mM D, L-glyceraldehyde.
4. 6 mM PGA.

5. 10 mM NaHCO$_3$, 6 mM PGA.
6. 10 mM malate.
7. 10 mM aspartate, 10 mM pyruvate.
8. 10 mM ribose 5-phosphate, 10 mM NaHCO$_3$, 10 mM malate.

*N.B. In some species, such as maize, the photosynthetic capacity of bundle-sheath strands is readily lost with prolonged enzymic digestion. For this reason, a rapid procedure combining enzymic digestion and mechanical separation yields the most active preparations (Chapman, Berry and Hatch, 1980).

CAM Cells and Protoplasts
$^{14}CO_2$ fixation and light-dependent O$_2$ evolution with and without bicarbonate may be measured in an assay medium containing 0.33 M sorbitol, 1 mM MgCl$_2$, 50 mM Tricine, pH 8.0.

6.5. CHLOROPLAST ISOLATION FROM PROTOPLASTS

The mesophyll protoplasts of C$_3$ and C$_4$ plants have an average diameter of 30–40 μm whilst those of CAM plants are at least 60–100 μm in diameter. A quick effective procedure for isolating chloroplasts is to pass the protoplasts several times through a 20-μm nylon mesh. This breaks all of the protoplasts but leaves chloroplasts and other smaller organelles largely intact. Breakage can be achieved by fitting nylon mesh to the end of a 1-ml disposable syringe which has had the tip excised to provide a pore of about 3 mm diameter. An aliquot of the protoplasts will be ruptured when taken up and ejected two or three times from the syringe. Intact chloroplasts (which have a diameter of 3 to 5 μm) will be released. (Although the above method is adequate and simple, other methods may be employed. For example, passing protoplasts through a Yeda Press at 75 psi will yield largely intact chloroplasts.) The protoplast extract can be centrifuged at low speed, e.g. 250 g for 1 minute, to obtain a chloroplast pellet largely free of the cytosol fraction. This is sufficient for many studies on photosynthesis with chloroplasts from various species. For studies on the intracellular localization of enzymes, the total protoplast extract can be layered on a sucrose density gradient and then centrifuged and fractionated by standard procedures.

Mesophyll protoplasts may also be broken by passing them through a 20-μm mesh and the centrifugation omitted. The chloroplasts are assayed in the presence of all the extra-chloroplastic enzymes and the preparation is referred to as protoplast extract.

C$_3$ Chloroplasts
Take an aliquot of protoplasts in sorbitol medium and centrifuge at 100 g for 5 minutes. Resuspend the protoplasts in the isolation, resuspension and assay medium containing 0.4 M sorbitol, 10 mM EDTA, 10 mM NaHCO$_3$, 25 mM Tricine-KOH pH 8.0. Break the protoplasts and centrifuge at 250 g for 60 seconds. Gently resuspend the chloroplast pellet and store on ice.

C$_4$ Chloroplasts
Chloroplasts from mesophyll protoplasts can be made in the same way as C$_3$ chloroplasts. The protoplasts should be resuspended in 0.33 M sorbitol, 1 mM MgCl$_2$, 1 mM MnCl$_2$, 1 mM EDTA, 2 mM KH$_2$PO$_4$, 50 mM Tricine-KOH, pH 7.5 and adjusted to about 200 μg chlorophyll/ml. This medium may be used for isolation, resuspension and assay. After isolation, chloroplasts should be washed once, stored on ice and assayed as soon as possible.

CAM Chloroplasts
For breakage of the protoplasts a 30 μm mesh rather than a 20-μm mesh should be used. For the breaking medium use 150 mM sorbitol, 1% (w/v) PVP (soluble), 5 mM EDTA, 5 mM Na$_4$P$_2$O$_7$, 250 mM Tricine-KOH, pH 8.2. A higher buffer concentration is used than with C$_3$ or C$_4$ protoplasts owing to the high acidity of CAM tissue. Centrifuge the protoplast extract at 100–200 g for 2 minutes. Resuspend the chloroplast pellet in 0.33 M sorbitol, 5 mM Na$_4$P$_2$O$_7$, 50 mM Tricine-KOH, pH 7.8.

6.6. THE MECHANICAL SEPARATION OF INTACT CHLOROPLASTS

At all steps of the isolation, glassware and solutions should be kept chilled.

Grinding Medium

Spinach

Any one of a number of sugars or sugar alcohols may be used as osmotica and any one of a number of buffers to maintain the pH. Some advantage is derived from the use of a slightly acid pH value and small quantities of $MgCl_2$ and EDTA are often included. However, "good" chloroplasts can be prepared in the absence of Mg (for example in 0.33 M sorbitol +10 mM pyrophosphate at pH 6.5) but the presence of Mg together with EDTA appears to be beneficial even though the reasons remain obscure.

The following grinding medium continues to be the most useful for many purposes: 0.33 M sorbitol, 10 mM $Na_4P_2O_7$, 5 mM $MgCl_2$, 2 mM Na isoascorbate, all adjusted to pH 6.5 with HCl.

The sodium ascorbate (or isoascorbate) should be added, as the salt, immediately prior to pH adjustment. The pyrophosphate should be freshly prepared (daily). The $MgCl_2$ is best added as an appropriate volume of a 1.0 M stock solution.

Peas

The above medium is less useful for peas because, with chloroplasts from this source, both ADP and PP_i inhibit when added singly but stimulate when added together. The solution first used for the isolated active pea chloroplasts is still as good as any for this purpose despite the high concentration of P_i, although a more inert buffer can be used at the outset or for washing: 0.33 M glucose, 50 mM Na_2HPO_4, 50 mM KH_2PO_4, 5 mM $MgCl_2$, 0.1% NaCl, 0.2% Na isoascorbate, 0.1% BSA, all adjusted to pH 6.5 with HCl.

Harvesting

Most leaves yield better chloroplasts if freshly harvested but spinach has been known to give good rates after 4 weeks of cold storage. Conversely pea shoots deteriorate very rapidly and should always be used immediately. If leaves are brightly illuminated for 20–30 minutes prior to grinding, chloroplast yield is increased and induction shortened.

Grinding

This is a compromise between opening as many cells and as few chloroplasts as possible; 3–5 seconds is appropriate for a conventional Waring blender (at full speed) and 3 seconds or less for the Polytron. The latter gives a higher yield. Freshly prepared media are best used as semi-frozen slush (solutions may be stirred in chilled alcohol at about −15° until a suitable consistency is achieved).

Filtration

The brei is usually squeezed through two layers of muslin (to remove coarse debris) and filtered through a sandwich of cotton wool between eight layers of muslin.

Centrifugation

Many variants may be used but it is important to separate the chloroplasts quickly from the supernatant. For example, if a brei of 200–300 ml is divided between four tubes of approximately 100 ml capacity in a swing-out head, adequate precipitation may be achieved by accelerating to approximately 6000 g and returning to rest in 90 seconds.

Resuspension

The supernatant is decanted and the pellet surface washed (e.g. with grinding or resuspending medium which is then discarded). This removes the upper layer of the pellet in which broken chloroplasts are more abundant. Small quantities of resuspending medium (0.5 ml/tube) are added and the chloroplasts resuspended by shaking the tubes or by using a small paint-brush or glass rod wrapped with cotton wool.

Resuspension and Assay Medium

This naturally depends on the nature of the investigation but the following, based on that of Jensen and Bassham, is useful for many purposes and (minus bicarbonate) is often also used as a resuspending medium.

0.33 M sorbitol, 2 mM EDTA, 1 mM $MgCl_2$, 1 mM $MnCl_2$, 50 mM HEPES, 10 mM $NaHCO_3$, 5 mM $Na_4P_2O_7$ (adjusted to pH 7.6 with KOH).

Note that phosphate must be included in the assay medium (see below).

6.7. PHOTOSYNTHESIS BY ISOLATED CHLOROPLASTS

Criteria of Intactness

Unless illuminated in mixtures containing additional ferredoxin, $NADP^+$, ADP, Mg, etc. (the constituents of the reconstituted system), envelope-free chloroplasts will not carry out CO_2-dependent O_2 evolution, CO_2-fixation, or PGA-dependent O_2 evolution. The best test of photosynthetic function is therefore simply to illuminate in near-saturating red light in an appropriate assay medium. At 20°C wheat or spinach chloroplasts which exhibit rates of CO_2-fixation or CO_2-dependent O_2 evolution of less than 50 mg^{-1} chlorophyll h^{-1} may be regarded as poor. Rates of 50–80 are reasonable, 80–150 are good and rates above 150 are exceptional.

Chloroplasts with intact envelopes (Class A) will not carry out many functions at fast rates because of the permeability barrier afforded by the inner envelope. They will not, for example:

1. Reduce exogenous oxidants such as ferricyanide or NADP.
2. Rapidly phosphorylate exogenous ADP.
3. Support O_2 uptake with non-permeating Mehler reagents such as ferredoxin.
4. Fix CO_2 in the dark with ribulose bisphosphate or ribose 5-phosphate + ATP as substrates.

5. Hydrolyse exogenous inorganic pyrophosphate.

All of these could be used as the basis of "intactness" assays. Possibly the simplest, and certainly that which has gained most favour, is the use of ferricyanide as Hill oxidant. The reduction of this compound may be measured spectrophotometrically or followed in the O_2 electrode provided that CO_2-dependent O_2 evolution is inhibited by glyceraldehyde. The assay probably overestimates intactness (there is evidence which suggests that rupture and loss of stromal protein may be followed by resealing) but it provides a useful basis for comparison. The response to the uncoupler is also useful. Well-coupled chloroplasts may show as much as a fourteen-fold increase in rate following the addition of NH_4Cl. Chloroplasts which show a four-fold response, or less, are unlikely to support carbon assimilation at good rates.

Intactness Assay

See Lilley et al., 1975

Chloroplast resuspension and assay medium:

0.33 M sorbitol,
1 mM $MgCl_2$,
1 mM $MnCl_2$,
2 mM EDTA,
50 mM HEPES-KOH, pH 7.6.

For osmotic shock of the chloroplasts, add 0.1 ml of chloroplast suspension to 0.9 ml of water and allow to stir for 1 minute. Add 1 ml of double-strength assay medium (0.66 M sorbitol, 2 mM $MgCl_2$, 2 mM $MnCl_2$, 4 mM EDTA, 100 mM HEPES-KOH, pH 7.6). For assay of the intact preparation, add 0.1 ml chloroplasts to 1.9 ml of assay medium. Chlorophyll should not exceed 100 μg/ml.

Add 10 mM D, L-glyceraldehyde to the assay (this inhibits the RPPP (reductive pentose phosphate pathway) and its associated O_2 evolution) and 3 mM potassium ferricyanide. Two or three minutes after illumination, add 5 mM NH_4Cl. Measure the rate of light-dependent O_2 evolution (μm O_2 evolved mg^{-1} Chl h^{-1}) after the addition of NH_4Cl with shocked (A)

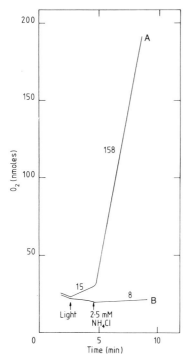

FIG. 6.2. Example of the test for intactness of chloroplasts using ferricyanide as a Hill oxidant. Chloroplasts were isolated from protoplasts of wheat. The calculated percentage intactness is 95%. Numbers along the traces indicate rates of O_2 evolution (μmol O_2 mg^{-1} Chl h^{-1}).

and unshocked (B) chloroplasts (see Fig. 6.2).

% intactness of original preparation =

$$\frac{A - B}{A}(100).$$

6.8. CARBON ASSIMILATION BY C₃ CHLOROPLASTS

The number of different media used for the assay of chloroplasts has increased considerably with the introduction of protoplasts. Their common features are the inclusion of an osmoticum (usually 0.33 M sorbitol); a buffer, HEPES or Tricine at pH 7.6–8.4; the presence of PP$_i$ or EDTA or a combination of both (the EDTA being

particularly useful with chloroplasts from protoplasts) and the inclusion of a saturating concentration of NaHCO₃ (10 mM) and orthophosphate.

Chloroplasts require P$_i$ for carbon assimilation because their major product is triose phosphate rather than free carbohydrate. Chloroplasts isolated from different tissues differ markedly in their P$_i$ optima. Whilst chloroplasts from spinach display maximum rates at 0.25–0.5 mM P$_i$ and frequently show a tolerance of high concentrations of P$_i$ (5–10 mM), chloroplasts from wheat have an optimum around 0.2 mM P$_i$ and are far less tolerant of high concentrations. In all experiments involving CO_2 fixation or CO_2-dependent O_2 evolution, the initial lag and the final rate will depend upon the balance between the levels of sugar phosphates within the chloroplast and the exogenous concentration of P$_i$. This dependency is a consequence of the operation of the phosphate translocator (Fliege *et al.*, 1978). The translocator (Fig. 6.3) brings about a rapid, obligatory exchange between P$_i$ and triose phosphate or

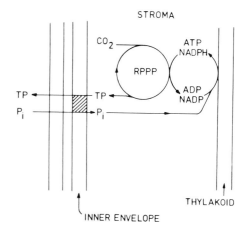

FIG. 6.3. The phosphate translocator, situated in the inner membrane of the chloroplast envelope, exchanges cytosolic phosphate for triose phosphate (TP) (comprising glyceraldehyde phosphate + dihydroxyacetone phosphate) and, to a lesser extent, for 3-PGA. Synthesis of sucrose, etc., in the cytosol recycles P$_i$ so that, in the steady state, there is a constant uptake of P$_i$ and export of TP by the chloroplast.

PGA (and to a lesser extent with some other sugar phosphates). Thus too much P_i is inhibitory to photosynthesis (e.g. Fig. 6.4b), especially during induction, because triose phosphate is forcibly exported from the chloroplast. The induction period or lag is due to the fact that when the light is switched on, the level of sugar phosphates within the chloroplast is low (Lilley *et al.*, 1977) but as CO_2 fixation proceeds, a proportion of the triose phosphate and PGA formed as the product is retained within the chloroplast, and through the operation of the RPPP, this results in an increase in the total amount of the CO_2 acceptor, ribulose bisphosphate. This allows a higher rate of photosynthesis and thus the generation of even more triose phosphate, and so on. The autocatalytic increase in total sugar phosphate reaches a ceiling imposed by other factors such as enzyme activity, etc.

For these reasons, PGA or triose phosphate added to the medium will overcome the initial lag (e.g. Fig. 6.4a). It is useful to examine PGA-dependent O_2 evolution (and/or PGA-dependent CO_2 fixation), especially with chloroplasts of suspect activity. If PGA-dependent O_2 evolution can be achieved with good rates, there is every reason to hope for good CO_2-dependent O_2 evolution following manipulation of P_i concentration, media, etc. 1 mM PGA usually gives rates between 60 and 100% of the rate of CO_2-dependent O_2 evolution.

The most useful medium for mechanically derived chloroplasts from spinach and peas is based on that of Jensen and Bassham (Section 6.6). The role of pyrophosphate is too complex to discuss in detail but several points are clear. Firstly, orthophosphate is less inhibitory to chloroplasts in the presence of PP_i because it is a competitive inhibitor of the phosphate translocator. Secondly, PP_i does not seem to cross the spinach chloroplast envelope, but in the presence of external Mg and pyrophosphatase (released from damaged chloroplasts) is slowly hydrolysed to P_i. It can thus act as an optimal source of P_i. With spinach chloroplasts the optimal concentration of P_i is about 0.25 mM if PP_i is not employed, but the optimum is so sharp that it is

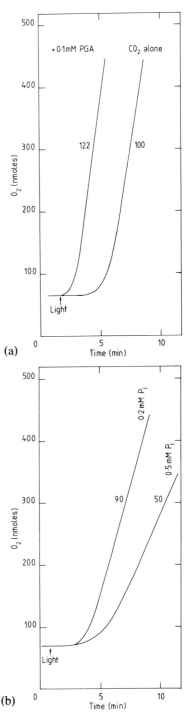

FIG. 6.4. (a) The shortening of the lag in CO_2-dependent O_2 evolution by (a) PGA and (b) its extension by super-optimal P_i in wheat chloroplasts. Numbers along the traces indicate rates of O_2 evolution (μ mol O_2 mg^{-1} Chl h^{-1}).

very difficult to achieve. Under such conditions the concentration of P_i will change rapidly during illumination unless the chlorophyll concentration is kept very low. Recent work with chloroplasts from young pea leaves and young wheat leaves suggests that ADP and ATP within the chloroplasts exchange with external PP_i. For this reason PP_i is inhibitory if used alone with pea or wheat chloroplasts but it stimulates photosynthesis in the presence of small quantities (0.2–0.4 mM) of ADP or ATP (Robinson and Wiskich, 1977; Edwards et al., 1978).

The most significant respects in which media for chloroplasts of wheat, etc., differ from that based on Jensen and Bassham are the omission of PP_i (for the reasons outlined above) and the inclusion of relatively high concentrations (10 mM) of EDTA (Section 6.5). Whilst the precise function of EDTA remains unclear, it seems that although it shares with PP_i the property of being a chelator of metal ions, its main function may be to provide a high concentration of negative charge in the medium. It should also be noted that wheat chloroplasts tend to have a more alkaline pH optimum for photosynthesis (pH 8.0–8.4) than do spinach chloroplasts.

Light-dependent O_2 evolution and $^{14}CO_2$ fixation may be measured using the same media. Chloroplasts are normally used at a concentration of 50–100 μg chl/ml in a volume of 1–2 ml. Above this concentration the light intensity may not be saturating. A 150 W quartz-iodine lamp (in a slide projector) together with a round-bottomed flask of water (which both focuses the light and absorbs heat), a red filter (e.g. ICI red perspex 400) and a Calflex C heat filter gives a light intensity of about 350 W m^{-2} (for details of the oxygen electrode apparatus see Delieu and Walker, 1972). The temperature employed for C_3 chloroplasts is normally 20°C, although it may be increased (up to 35°C) for material from C_4 plants.

It may be beneficial to add catalase to the assay medium (100 units/ml). This prevents H_2O_2 accumulation (from pseudocyclic electron flow) which inhibits chloroplast photosynthesis.

To determine the phosphate optimum, do separate assays with varying concentrations of P_i, e.g. 0, 0.1, 0.3, 0.5, 1.0, 5.0 and 10.0 mM.

Other assays which can be tried:

1. 10 mM NaHCO$_3$, 0.3 mM P_i (or optimal P_i);
2. 10 mM NaHCO$_3$, 0.3 mM P_i, 10 mM D, L-glyceraldehyde;
3. 10 mM NaHCO$_3$, 0.3 mM P_i, 1 mM pyridoxal phosphate;
4. 1 mM PGA;
5. 1 mM PGA, 10 mM D, L-glyceraldehyde;
6. 5 mM PGA;
7. 1 mM ribose 5-phosphate;
8. 1 mM fructose 6-phosphate.

Note that the chloroplast is not solely concerned with carbon fixation, and that it will reduce compounds involved in nitrogen metabolism (Section 8) and will metabolize many other compounds. As an example, oxaloacetate enters the chloroplast on the dicarboxylate translocator and is reduced to malate by NADP-malate dehydrogenase. This can be seen by adding 5 mM oxaloacetate (freshly made-up solution, pH 7–8) to a suspension of illuminated chloroplasts in the absence of bicarbonate. After 3 or 4 minutes, add 5 mM NH$_4$Cl to uncouple the chloroplasts.

C$_4$ Mesophyll Chloroplasts

O_2 evolution: assay the chloroplasts in 0.33 M sorbitol, 1 mM MgCl$_2$, 1 mM MnCl$_2$, 1 mM EDTA, 2 mM KH$_2$PO$_4$, 50 mM Tricine-KOH, pH 7.5, in the oxygen electrode, using an assay volume of 1–2 ml, with about 20 μg Chl/ml. Compare protoplast extracts and chloroplasts using the following substrates:

1. 2 mM PGA;
2. 0.5 mM OAA + 5 mM pyruvate after 3 minutes of illumination;
3. 0.5 mM OAA + 5 mM NH$_4$Cl after 3 minutes of illumination;
4. 10 mM NaHCO$_3$;
5. 5 mM pyruvate;
6. 5 mM pyruvate + 2 mM KCN (which inhibits catalase);
7. 5 mM pyruvate + 200 units of catalase;
8. 5 mM ribose + 5 mM phosphate;
9. 1 mM p-benzoquinone;
10. 5 mM NH$_4$Cl.

Which substrates appear to serve as natural Hill oxidants with C_4 mesophyll chloroplasts? Do any of the results suggest potential for pseudo-cylic photophosphorylation?

CO$_2$ *fixation*: $^{14}CO_2$ fixation studies can be done with mesophyll protoplast extracts and chloroplasts exactly as described for protoplasts (Section 6.4).

CAM Chloroplasts

Compare CO_2 fixation in chloroplasts and protoplasts in the light and dark, following the method described in Section 6.4. A suitable assay medium for chloroplasts includes 330 mM sorbitol, 5 mM EDTA, 5 mM Na$_4$P$_2$O$_7$, 50 mM Tricine-KOH, pH 7.6 and 100 units of catalase/ml.

6.9. MATERIALS

Examples of sources of some of the special material which is used for protoplast isolation. Note that cellulase and pectinase stored at −20°C retain most of their activity for at least a year.

Cellulase from *Trichoderma viride*
As Onozuka 3S or R10
 Yakult Biochemical Co., Ltd.,
 Enzyme Products,
 8–21 Shingikancho,
 Nishinomiyo,
 Japan

As Cellulysin Calbiochem, Ltd.,
 C.P. Laboratories Ltd.,
 P.O. Box 22,
 Bishops Stortford,
 Herts. CM23 3AL,
 England

 Calbiochem, Ltd.,
 P.O. Box 12087,
 San Diego, CA 92112,
 U.S.A.

Pectinase from *Rhizopus* sp.
As Macerozyme R10
 Yakult Biochemical Co., Ltd.
 (see above)

As Macerase
 Calbiochem, Ltd.
 (see above)

Pectinase from *Aspergillus* sp.
As Extractase PC
 Fermco Biochemics, Inc.,
 2638 Delta Lane,
 Ik Grove Village, IL 60007,
 U.S.A.

As Rohament P
 Rohm GmbH Chemische Fabrik,
 D-1600 Darmstadt,
 Kirschenallee,
 Postfach 4242,
 Germany

As Pectolyase Y-23
 Seishin Pharmaceutical Co. Ltd.,
 9-500-1, Nagareyama,
 Nagareyama-shi,
 Chiba-ken,
 Japan

Nylon mesh Henry Simon, Ltd.,
 P.O. Box 31,
 Stockport,
 Cheshire, SK3 0RT,
 England

 Tetko, Inc.,
 Precision Woven
  Screening Media,
 420 Saw Mill River Road,
 Elmsford, NY 10523,
 U.S.A.

Dextran T$_{20}$ U.S. Biochemicals Corp.,
 21000 Miles Parkway,
 Cleveland, OH 44128,
 U.S.A.

Bibliography and Further Reading

General

EDWARDS, G. E. and WALKER, D. A. C3, C4: Some aspects of photosynthetic carbon assimilation. (Packard Publishing Co. Ltd.: Chichester.) (In press.)

Chlorophyll Determination

ARNON, D. I. (1949) Copper enzymes in isolated chloroplasts. Polyphenoloxidase in *Beta vulgaris*. *Plant Physiol.* **24**: 1–15.

WINTERMANS, J. F. G. M. and DE MOTS, A. (1965) Spectrophotometric characteristics of chlorophylls a and b and their phenophytins in ethanol. *Biochim. Biophys. Acta* **109**: 448–453.

Protoplast and Chloroplast Isolation

EDWARDS, G. E., HUBER, S. C. and GUTIERREZ, M. (1976) Photosynthetic properties of plant protoplasts. In: *Microbial and Plant Protoplasts*, pp. 229–332 (eds. J. F. PEBERDY, A. H. ROSE, H. J. ROGERS and E. C. COCKING. Academic Press, New York.)

EDWARDS, G. E., ROBINSON, S. P., TYLER, N. J. C. and WALKER D. A. (1978) Photosynthesis by isolated protoplasts, protoplast extracts, and chloroplasts of wheat. Influence of orthophosphate, pyrophosphate, and adenylates. *Plant Physiol.* **62**: 313–317.

EDWARDS, G. E., LILLEY, R. McC., CRAIG, S. and HATCH, M. D. (1979) Isolation of intact and functional chloroplasts from mesophyll and bundle sheath protoplasts of the C4 plant *Panicum miliaceum*. *Plant Physiol.* **63**: 821–827.

HALL, D. O. (1972) Nomenclature for isolated chloroplasts. *Nature New Biol.* **235**: 125–126.

HUBER, S. C. and EDWARDS, G. E. (1975) An evaluation of some parameters required for the enzymatic isolation of cells and protoplasts with CO_2 fixation capacity from C3 and C4 grasses. *Physiol. Plant* **35**: 203–209.

KANAI, R. and EDWARDS, G. E. (1973) Purification of enzymatically isolated mesophyll protoplasts from C3, C4, and Crassulacean Acid Metabolism plants using an aqueous dextran polyethylene glycol two-phase system. *Plant Physiol.* **52**: 484–490.

LEEGOOD, R. C. and WALKER, D. A. (1979) Isolation of protoplasts and chloroplasts from flag leaves of *Triticum aestivum* L. *Plant Physiol.* **63**: 1212–1214.

ROBINSON, S. P. and WALKER, D. A. (1979) Rapid separation of the chloroplast and cytoplasmic fractions from intact leaf protoplasts. *Arch. Biochem. Biophys.* **196**: 319–323.

WALKER, D. A. (1971) Chloroplasts (and grana) – aqueous (including high carbon fixation ability). In: *Methods in Enzymology* (ed. A. SAN PIETRO), **23**: 211–220 (Academic Press, New York.)

Isolated Chloroplasts

COCKBURN, W., BALDRY, C. W. and WALKER, D. A. (1967) Some effects of inorganic phosphate on O_2 evolution by isolated chloroplasts. *Biochim. Biophys. Acta* **143**: 614–624.

DELIEU, T. and WALKER, D. A. (1972) An improved cathode for the measurement of photosynthetic oxygen evolution by isolated chloroplasts. *New Phytol.* **81**: 201–225.

FLIEGE, R., FLUGGE, U-I., WERDAN, K. and HELDT, H. W. (1978) Specific transport of inorganic phosphate, 3-phosphoglycerate and triosephosphates across the inner membrane of the envelope in spinach chloroplasts. *Biochim. Biophys. Acta* **502**: 232–247.

LILLEY, R. McC. and WALKER, D. A. (1979) Studies with the reconstituted chloroplast system. In: *Encyclopedia of Plant Physiology*, New Series (Eds. M. GIBBS and E. LATZKO), **6**: 41–53. (Springer-Verlag, Berlin.)

LILLEY, R. McC., FITZGERALD, M. P., RIENITS, K. G. and WALKER, D. A. (1975) Criteria of intactness and the photosynthetic activity of spinach chloroplast preparations. *New Phytol.* **75**: 1–10.

LILLEY, R. McC., CHON, C. J., MOSBACH, A. and HELDT, H. W. (1977) The distribution of metabolites between spinach chloroplasts and medium during photosynthesis *in vitro*. *Biochim. Biophys. Acta* **460**: 259–272.

ROBINSON, S. P. and WISKICH, J. T. (1977) Pyrophosphate inhibition of carbon dioxide fixation in isolated pea chloroplasts by uptake in exchange for endogenous adenine nucleotides. *Plant Physiol.* **59**: 422–427.

C4 Photosynthesis

CHAPMAN, K. S. R., BERRY, J. A. and HATCH, M. D. (1980) Photosynthetic metabolism in bundle sheath cells of the C4 species *Zea mays*: Sources of ATP and NADPH and the contribution of photosystem II. *Arch. Biochem. Biophys.* **202**: 330–341.

EDWARDS, G. E., HUBER, S. C., KU, S. B., RATHNAM, C. K. M., GUTIERREZ, M., and MAYNE, B. C. (1976) Variation in photochemical activity in C4 plants in relation to CO_2 fixation. In: CO_2 *Metabolism and Productivity of Plants* (Eds. R. H. BURRIS and C. C. BLACK), pp. 83–112. (University Park Press, Baltimore, MD.)

HUBER, S. C. and EDWARDS, G. E. (1975) C4 photosynthesis. Light dependent CO_2 fixation by mesophyll cells, protoplasts, and protoplast extracts of *Digitaria sanguinalis*. *Plant Physiol.* **55**: 835–844.

HUBER, S. C. and EDWARDS, G. E. (1975) Effect of DBMIB, DCMU, and antimycin A on cyclic and noncyclic electron flow in C4 mesophyll chloroplasts. *FEBS Lett.* **58**: 211–214.

HUBER, S. C. and EDWARDS, G. E. (1975) The effect of oxygen on CO_2 fixation by mesophyll protoplast extracts of C3 and C4 plants. *Biochem. Biophys. Res. Commun.* **67**: 28–34.

Huber, S. C. and Edwards, G. E. (1976) Studies on the path of cyclic electron flow in mesophyll chloroplasts of a C₄ plant. *Biochim. Biophys. Acta* **449**: 420–433.

Crassulacean Acid Metabolism

Spalding, M. H. and Edwards, G. E. (1978) Photosynthesis in enzymatically isolated leaf cells from the CAM plant *Sedum telephium* L. *Planta* **141**: 59–63.

Spalding, M. H. and Edwards, G. E. (1980) Photosynthesis in isolated chloroplasts of the Crassulacean acid metabolism plant *Sedum praealtum*. *Plant Physiol.* **65**: 1044–1048.

Spalding, M. H., Schmitt, M. R., Ku, S. B., and Edwards, G. E. (1979) Intracellular localization of some key enzymes of Crassulacean acid metabolism in *Sedum praealtum*. *Plant Physiol.* **63**: 738–743.

APPENDIX A6. OXYGEN ELECTRODE

An oxygen electrode (e.g. Fig. 6.7) is a special form of electrochemical cell, in which a current is generated that is proportional to the activity of oxygen present in a solution. In principle, the electrode consists of two wires, one made of platinum and one of silver coated with silver chloride. These wires dip into a solution of electrolyte, e.g. KCl. An electrical potential is applied across the wires, with the platinum electrode being made negative with respect to the silver, and this causes the following electrochemical reactions to occur:

At the silver electrode:

$$4\,Ag \leftrightharpoons 4\,Ag^+ + 4\,e^-$$

At the platinum electrode:

$$O_2 + 2\,e^- + 2H^+ \leftrightharpoons (H_2O_2)$$
$$(H_2O_2) + 2\,e^- + 2H^+ \leftrightharpoons 2H_2O$$

If the wires are connected to a battery a current will flow from the silver to the platinum, and under appropriate conditions, the magnitude of the current will be linearly proportional to the oxygen activity in the solution. If no oxygen is present the reaction at the platinum electrode will not occur and no current should flow.

The electrode as described is inconvenient because the platinum can become poisoned by biological samples and then will not respond to oxygen tension.

The platinum and reference electrodes are shielded from the solution by a thin membrane which is permeable to oxygen but impermeable to most poisons for platinum. The membrane is kept in place by a rubber "O" ring. The platinum disc electrode consumes oxygen and to maintain a stable oxygen gradient across the membrane it is necessary to stir the solution at about 500–600 rpm with a small magnetic follower.

Polarizing Voltage

It is necessary to apply a polarizing voltage to the oxygen electrode to cause a current to flow. If the electrode output at a given oxygen activity is plotted against polarizing voltage, it is found that a plateau exists between about 0.4–0.8 volt (platinum negative with respect to the reference electrode). If a polarizing voltage near the centre of the plateau is chosen (-0.65 volt) then the current output from the oxygen electrode is linearly proportional to the oxygen tension of the solution.

The Electrode Reactions (Fig. 6.5)

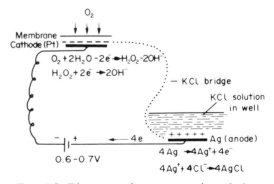

FIG. 6.5. Diagrammatic representation of electrode reactions. When the potentiating voltage is applied across the two electrodes the platinum becomes negative and the silver becomes positive. Oxygen diffusing through the membrane is reduced at the platinum surface and a current flows through the circuit (which is completed by the KCl bridge). The silver is oxidized and silver chloride is deposited. The current is stoichiometrically related to the oxygen reduced.

A Potentiating Circuit for a Double Electrode Set-up (Fig. 6.6)

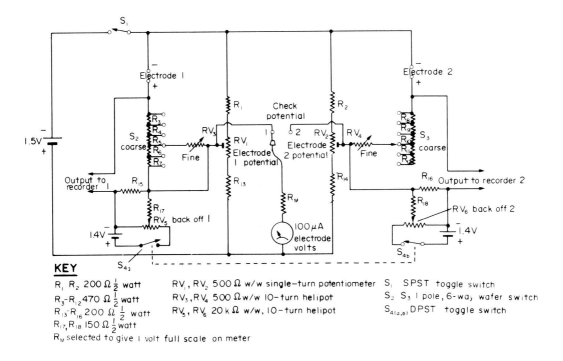

KEY

R_1, R_2 200 Ω $\frac{1}{2}$ watt	RV_1, RV_2 500 Ω w/w single–turn potentiometer	S_1 SPST toggle switch
R_3-R_{12} 470 Ω $\frac{1}{2}$ watt	RV_3, RV_4 500 Ω w/w 10–turn helipot	$S_2 \; S_3$ I pole, 6-way wafer switch
R_{13}-R_{16} 200 Ω $\frac{1}{2}$ watt	RV_5, RV_6 20 k Ω w/w, 10–turn helipot	$S_{4(a,b)}$ DPST toggle switch
R_{17}, R_{18} 150 Ω $\frac{1}{2}$ watt		
R_M selected to give I volt full scale on meter		

FIG. 6.6. Potentiating and zero suppress circuit. This circuit is used to apply a potentiating voltage of 600–700 mV simultaneously across the electrodes of each of two electrode vessels. The variable resistors RV_1 and RV_2 are used to adjust this voltage using the meter and the wander plug in socket 1 or 2 as appropriate. The variable resistors RV_3 (fine) and stepped fixed resistors ($S_2 + R_{3-7}$) are then used to adjust the output from electrode 1 to the recorder to approximately 1 mV and RV_4 and $S_3 + R_{8-12}$ fulfil a similar function for electrode 2. RV_3 and RV_4 are then used to set the air-line at an appropriate exact value during calibration. RV_5 and RV_6 are used to offset the air-lines prior to the start of measurement if oxygen evolution is being followed and the available space on the chart is insufficient for this purpose.

Characteristics of the Oxygen Electrode

1. Membranes

Many different types of membrane have been tested including collodion, cellophane, polythene, nylon, "Teflon", "Mylar", silicone rubber and "Cling-film". Oxygen permeability of the membrane will depend on both composition and thickness, and the thinnest membrane that can reasonably be used is about 0.0005 in. The most commonly used membranes are polythene and Teflon in thicknesses of 0.0005–0.001 in. (12.5–25 μm). Thinner membranes will give increased response speed at the expense of increased fragility.

2. Output Current

The current generated by an oxygen electrode is governed by the amount of oxygen reaching the platinum cathode, and by the cathode area. Thus for a given oxygen tension and membrane, a large platinum disc will give a large output current, but will consume measurable amounts of oxygen from the solution. A large cathode will require better stirring to maintain equilibrium of oxygen between the solution and the electrode, compared with a smaller cathode. For a given set of conditions and membrane type, a small cathode will give a faster response time than a large cathode.

3. Response Time

The 90% response time of an oxygen electrode is defined as the length of time required for the electrode to make a 90% response to an instantaneous change in oxygen tension in the solution. Factors affecting the response time have already been mentioned, namely the thinness and composition of the membrane, the size of the cathode and the rate of stirring.

Using modern electrode designs and an 0.5–1.0-mm cathode, a response time of 1–10 seconds is normal at 25°C. A very rough check on the response time can be made by switching off the stirrer when the electrode output will fall because oxygen diffusion across the membrane

is not fast enough to balance the oxygen reduction at the cathode. After the output has fallen by about three recorder units, the stirrer can be switched back on at the same speed as before, and the time taken to recover 90% of the former signal can be measured.

4. Electrode Leakage Current

In theory, a perfect electrode will give zero output when the oxygen tension in the solution is zero. In practice, small faults in the sealing of the platinum electrode to the insulator can occur and these will cause some leakage current to flow in the absence of oxygen. This can be tested for by setting up the electrode to read 100% on air-saturated water, and then adding a few crystals of sodium hydrosulphite (dithionite) which will chemically remove all the oxygen in the solution. Any remaining signal will then be caused by electrode leakage cur-

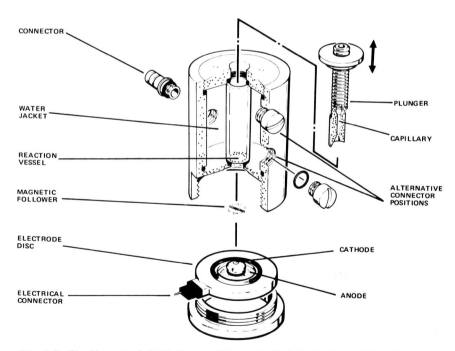

FIG. 6.7. The Hansatech D.W. Oxygen Electrode Unit is a high-precision laboratory instrument ideally suited for measuring the kinetics of oxygen uptake or evolution. It is a Clark type O_2 electrode based on a design by Delieu and Walker. (*New Phytologist* (1972) **71**, 201–225.) Obtainable from Hansatech Ltd., Paxman Road, Hardwick Industrial Estate, King's Lynn, Norfolk, England. A similar electrode is also made by Rank Brothers, High Street, Bottisham, Cambridge, CB5 9DA, England.

rent. It is advisable to test electrodes regularly in this way to detect the appearance of leakage currents.

5. Exclusion of Oxygen

The oxygen electrode measures the rate of oxygen uptake by measuring the disappearance of oxygen from the solution. It is essential for this purpose that oxygen cannot enter the reaction vessel at an appreciable rate. This is particularly important when monitoring reactions where the oxygen concentration changes slowly. Atmospheric oxygen is normally excluded from the reaction vessel by a close-fitting plastic disc which is drilled with a small hole (1 mm) through which additions can be made with an "auto-zero" type pipette or a Hamilton syringe. It is essential that no bubbles are allowed to remain under the disc when the electrode is in use, as this will cause rapid reoxygenation of the medium. The liquid level in the chamber should reach the top of the covering disc.

Calibration of the Oxygen Electrode

1. Direct calibration

It is normal to fill the oxygen electrode with air-saturated buffer and then use the gain control to set the meter to 100% with the electrode plugged in. It should be noted that a marked drift of output occurs for the first 2 or 3 minutes after the electrode is plugged in, while oxygen equilibration takes place. The amount of oxygen present in air-saturated solution may now be estimated by observing the oxidation of a known amount of substrate in the presence of an oxidase enzyme. One example of this system would be the use of submitochondrial particles to oxidize a known amount (i.e. $1 \mu M$) of NADH, previously standardized by spectrophotometric means.

The solubility of oxygen in water at various temperatures and pressures is shown in the following tables.

Solubility of Oxygen in Water (Saturated with Air) in PPM at Various Temperatures and Pressures

P (mm)	775	760	750	725	700	675	650	625
P (in.)	30.51	29.92	29.53	28.54	27.56	26.57	25.59	24.61
15(°C)	10.4	10.2	10.0	9.7	9.3	9.0	8.7	8.3
16	10.1	9.9	9.8	9.5	9.1	8.8	8.5	8.1
17	9.9	9.7	9.6	9.3	9.0	8.6	8.3	8.0
18	9.7	9.5	9.4	9.1	8.8	8.4	8.1	7.8
19	9.5	9.3	9.2	8.9	8.6	8.3	8.0	7.6
20	9.3	9.2	9.1	8.7	8.4	8.1	7.8	7.5
21	9.2	9.0	8.9	8.6	8.3	8.0	7.7	7.4
22	9.0	8.8	8.7	8.4	8.1	7.8	7.5	7.2
23	8.8	8.7	8.5	8.2	8.0	7.7	7.4	7.1
24	8.7	8.5	8.4	8.1	7.8	7.5	7.2	7.0
25	8.5	8.4	8.3	8.0	7.7	7.4	7.1	6.8
26	8.4	8.2	8.1	7.8	7.6	7.3	7.0	6.7
27	8.2	8.1	8.0	7.7	7.4	7.1	6.9	6.6
28	8.1	7.9	7.8	7.6	7.3	7.0	6.7	6.5
29	7.9	7.8	7.7	7.4	7.2	6.9	6.6	6.4
30	7.8	7.7	7.6	7.3	7.0	6.8	6.5	6.2
31	7.7	7.5	7.4	7.2	6.9	6.7	6.4	6.1
32	7.6	7.4	7.3	7.0	6.8	6.6	6.3	6.0

Temp. (°C)	O_2 (ppm)	O_2 (μm/ml)
0	14.16	0.442
5	12.37	0.386
10	10.92	0.341
15	9.76	0.305
20	8.84	0.276
25	8.11	0.253
30	7.52	0.230
35	7.02	0.210

Relation of Atmospheric Pressure to Altitude

Altitude	True atmospheric pressure (mm)
Sea level	760
1000 feet	733
2000 feet	707
3000 feet	681
4000 feet	656
5000 feet	632

SECTION 7

PHOTOSYNTHETIC ENERGY CONVERSION

by G. HIND

The function of the photosynthetic electron transport pathway shown in Fig. 7.1 is to convert light energy into chemical energy which is stored as reductant in the form of NADPH and as phosphorylation potential in the form of ATP. These products and their precursors are present in only catalytic amounts in the intact chloroplast, so oxygen evolution—a convenient indicator of the electron transport rate—is normally limited by the rate at which fixation of CO_2 regenerates $NADP^+$ and ADP.

The link between electron transport and phosphorylation is a gradient of pH across the thylakoid membrane. The gradient arises from H^+ ions ("protons") deposited by oxidation of water on the inner face of the membrane, and by passage of electrons through interacting hydrogen ($H^+ + e^-$) and electron carriers that are oriented in the membrane so as to move protons inwards, as shown in Fig. 7.1. A pH difference of at least 3.0 units ($[H^+]_{in}/[H^+]_{out} = 1000/1$) arises in this fashion. Protons move *out* of the thylakoid via the coupling factor (CF) proteolipid which consists of a proton channel (CF_0) spanning the membrane and a reversible ATPase (CF_1) that protrudes from the outer face of the thylakoid into the stroma. The CF_1 ATPase activity is optimal around pH 8.2, where the condensation of ADP and orthophosphate (P_i) requires removal of a hydroxyl ion (OH^-). Somehow, CF_1 uses the proton gradient to drive the following process:

$$ADP^{3-} + P_i^{2-} + H^+ = ATP^{4-} + H_2O$$

$$(pH > 8, +Mg^{2+})$$

When ADP is not available to CF_1, the proton gradient cannot be used and its backpressure on electron transport inhibits CO_2 fixation.

The inward proton movement driven by electron transport is linked to efflux of Mg^{2+} ions from the thylakoid. In the light then, the stroma experiences a shift to more alkaline pH and an increase in Mg^{2+} activity which, in turn, bring about activation of key enzymes in the carbon-reduction pathway. Through these and other processes, the soluble enzymes are regulated by electron transport in the thylakoid membrane.

The intact chloroplast can be manipulated to give high rates of oxygen evolution in the absence of CO_2 by providing an alternative terminal electron acceptor such as oxaloacetate together with an uncoupler. Uncouplers are a wide range of compounds that either buffer the internal protons or cause them to leak back across the thylakoid membrane so that a proton gradient does not develop; they thus prevent inhibitory backpressure of the proton gradient on electron flow.

If the chloroplast outer envelope is ruptured by osmotic shock, the stroma contents are lost and CO_2 fixation cannot occur unless these are reconstituted. However, the absence of the permeability barrier presented by the envelope makes possible the use of a wide range of unnatural terminal oxidants, such as the ferricyanide ion, and the addition of substrate amounts of ADP. Most of the early studies of electron flow and phosphorylation were done with such preparations of broken chloroplasts.

Careful rupture of chloroplasts in the presence of $MgCl_2$ gives a preparation which will phosphorylate ADP even though $NADP^+$ or artificial acceptors are lacking. This

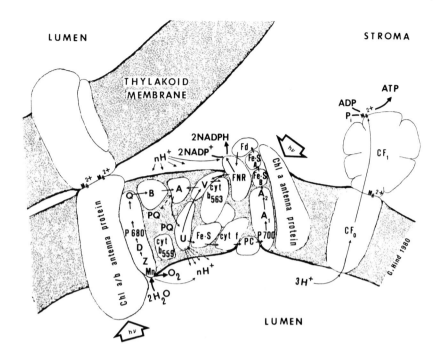

FIG. 7.1. Diagrammatic cross-section of thylakoid membranes, showing electron transfer components. On the left, two membranes are linked to form part of a grana stack by interaction of Mg^{2+} with charges on the surface of chlorophyll a/b pigment protein complexes. The stromal portion of the membrane, on the right, is deficient in these chlorophyll a/b complexes.

Z and D are unknown electron donors to the reaction centre (P680) of photosystem II; Q is the fluorescence-quenching carrier. B, A, U and V are probably all specialized protein-bound plastoquinones, with A being in equilibrium with the bulk plastoquinone pool (PQ). The high potential non-haem iron–sulphur protein (Fe·S) mediates electron flow between the plastoquinone A pool and cytochrome f. This in turn is oxidized by the copper protein plastocyanin (PC) (or in some algae, by a low molecular weight c-type cytochrome) which feeds electrons to the reaction centre of photosystem I (P700).

Intermediates $A_{1,2}$ in photosystem I are very short lived: A_1 is a dimer of chlorophyll a and A_2 is an iron–sulphur centre. The bound iron–sulphur centres A and B have very low potentials and reduce soluble ferredoxin (Fd) associated with the outer surface of the thylakoid in a complex with ferredoxin:NADP reductase (FNR). This enzyme either reduces $NADP^+$ (non-cyclic electron flow) or transfers electrons to V (cyclic electron flow). Cytochrome b_{563} is associated with V, probably as part of a two-electron gate involved in reducing U via A. Oxidation of U at an appropriate ambient redox potential results in passage of one of its electrons to the iron–sulphur centre (Fe·S) and one to V. Movement of the charge between U and V can be detected optically as an electrochromic effect (P518) much slower than that resulting from movement of charge through the photosystems. Potentials do not develop from the reduction of U by A because a proton moves together with the electron.

Operation of the coupling factor complex is discussed in the text. Note that Mg^{2+} is involved in the binding of CF_1 to CF_0 as well as in thylakoid stacking and (not shown) the complexing of Fd to FNR. Interactions shown for b-type cytochromes and for Fe · S centres are speculative.

phosphorylation is driven by cyclic flow of electrons around photosystem I (see Fig. 7.1) and in intact chloroplasts it contributes ATP needed for CO_2 fixation. In its sensitivity to inhibitors and the ambient reduction/oxidation (redox) potential, cyclic electron flow in higher plants resembles electron transport in primitive photosynthetic bacteria.

Study of the Proton Flux and Photophosphorylation Using a pH Electrode

Technical Aspects

Refer first to the appendix of this section for the theory of pH measurement. Light-induced proton fluxes give rise to only small pH changes in the suspension because the ratio of thylakoid volume to suspending medium volume is very small, hence some amplification of the signal from the pH meter is needed. Commercial pH meters, even those with an expanded scale feature, do not provide adequate gain and have insufficient offset to suppress the output for presentation on the sensitive ranges of a chart recorder. The simple amplifier shown in Fig. 7.2 provides offset and gain and is cheap to construct.

A stirred, thermostatted reaction vessel which can be made from glass is shown in Fig. 7.3. It is

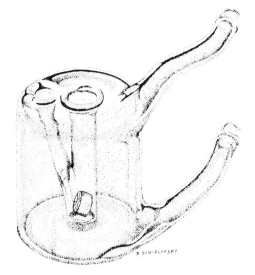

FIG. 7.3. A glass reaction vessel for measurement of proton flux using a pH electrode.

designed to accommodate a thin combination pH electrode and about 2 ml of sample. The side arm is used for making additions and cleaning the chamber without removing the electrode. A suction line connected to a length of thin polyethylene tubing facilitates cleaning.

Experiments

If the chloroplast preparation has high initial intactness, cyclic electron flow will drive the

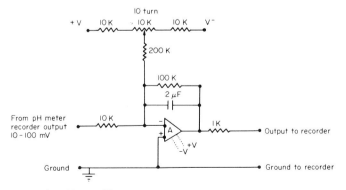

FIG. 7.2. pH amplifier and offset circuit. A is any cheap operational amplifier, e.g. LF 356, LM 741C, LM 709, MC 748, MC 1439, μA 709, μA 741C; V is 12 to 15 V d.c. from solid state power supply, or $+/-6$ V from ER 509 lantern batteries connected in series and centre-tapped to ground. The recorder should have ten-fold higher sensitivity (1 to 10 mV) than the pH meter output.

reactions under study and this endogenous system should be used. If the observed changes are small and noisy, however, an artificial electron acceptor should be provided. After measuring the control rate without acceptor, test the effect of 50 μM pyocyanine or of 50 μM methyl viologen plus 200 units of catalase and use one of these systems if necessary.

(A) Measurement of Proton Uptake

Intact chloroplasts are freshly shocked in the reaction vessel. Add 0.9 ml of 10 mM $MgCl_2$ containing 1 mM HEPES buffer, pH 6.5, and while slowly stirring, add 0.1 ml of chloroplasts (approx. 1 mg chlorophyll/ml). After 2 minutes (to allow for osmotic rupture) add 1.0 ml of 0.66 M sorbitol and adjust the pH to 6.5 ± 0.1 by cautious addition of 10 mM NaOH or HCl. Illuminate with strong light from a slide projector, focused by means of a water-filled round-bottom flask (250 ml) and filtered through red Perspex, until no further pH change occurs. Observe the effect of the reagents listed below on (1) the extent of the change and (2) the rate of relaxation after turning the light off, given as a half-time. Compare the effect of addition prior to illumination with addition during the illuminated steady state. Calibrate the proton uptake by addition of known amounts of 10 mM NaOH. Express the extent on a chlorophyll basis: microequivalents H^+ taken up per micromole chlorophyll. Express the uptake rate (net initial slope) as microequivalents H^+/(mg chlorophyll $\times$ hr).

(B) Measurement of Photophosphorylation

Proceed as in (A) but after osmotic rupture add 1 ml of RMP $\times$ 2 (see below) followed by 10 μl of 0.1 M ADP (pH 7.5–8.2). After adjusting the pH to 8.2 with 50 mM NaOH, measure the linear rate of alkalization in the light. Calibrate by adding known amounts of 50 mM HCl and from the equation calculate the ATP formed. Express results as micromoles ATP/(mg chlorophyll $\times$ hr). Reagents to be tested in (A) and (B) (concs. are final values): NH_4Cl 1 mM, carbonyl cyanide m-chlorophenylhydrazone (CCCP) 1 μM.

Compare traces in which the extent of the pH change is inhibited approximately 50% by (i) CCCP and (ii) reducing the light intensity with the aid of wire screens. Pay particular attention to the initial slopes of the pH rise and the half-times for relaxation.

RMP $\times$ 2:
Sorbitol 0.66 M, Tris 1 mM and Na^+ phosphate 2 mM (pH 8.2), KCl 20 mM.

Bibliography and Further Reading

Methodology

NISHIMURA, N., ITO, T. and CHANCE, B. (1962) *Biochimica et Biophysica Acta* **59**, 177–182.

How Cells Make ATP

HINKLE, P. C. and McCARTY, R. E. (1978) *Scientific American* **238**, 104–123.

Appendix A.7. pH Measurement

The Concept of pH

Water dissociates according to the reaction:

$$2H_2O \rightleftharpoons H_3O^+ + OH^-$$

The H_3O^+ ion is called an hydronium ion, or hydrated proton. Pure water dissociates only to a very small extent such that the concentration of H_3O^+ and OH^- are both 10^{-7} M at 25°C. The equilibrium constant for this dissociation is given by:

$$K_w = \frac{a\,H_3O^+ \times a\,OH^-}{a\,H_2O^2}$$

where a is the activity of the ion in solution.

Since pure water is 55.5 molar, and only a very small amount is ionized (10^{-7} M), water can be considered to be in its standard state and hence to have an activity of unity. Thus the above expression for the equilibrium constant becomes:

$$K_w = a\,H_3O^+ \times a\,OH^- = (H_3O^+)(OH^-)$$
$$= 10^{-7} \times 10^{-7} = 10^{-14}.$$

It becomes awkward to handle very small negative powers of ten to express proton concentrations and for this reason it is convenient to define a unit called pH which is the negative logarithm of the proton concentration.

$$pH = -\log_{10}(H_3O^+).$$

Strictly pH is $-\log_{10} H_3O^+$ activity, but for dilute aqueous solutions activity and concentration are very similar. Hence the pH of pure water (where $(H_3O^+) = 10^{-7}$) is 7.0.

Acids and Bases

The definition of acids and bases which is most often used in biological sciences is that formulated by Bronsted and Lowry, where an acid is defined as a potential proton donor and a base as a potential proton acceptor.

An acid will dissociate to produce a proton (which will be subsequently hydrated in aqueous solution) and a conjugate base according to:

$$HA + H_2O \overset{K_a}{\rightleftharpoons} H_3O^+ + A^-$$

where HA is the acid, A^- is the conjugate base, and K_a is the acid dissociation constant.

The acid dissociation constant K_a is a measure of the tendency that an acid has to donate a proton. A strong acid, which dissociates more readily than a weak acid, therefore has a high dissociation constant.

$$K_a = \frac{(A^-)(H_3O^+)}{(HA)}.$$

pK_a is defined as the negative logarithm of the acid dissociation constant and is used for the same reasons of convenience as apply in the case of pH:

$$pK_a = -\log_{10} K_a.$$

Inspection of the above expression for K_a will make clear that when an acid is half dissociated and $(HA) = (A^-)$ the value K_a will be equal to the H_3O^+ concentration. It follows that when an acid is half-dissociated $pH = pK_a$.

In a similar fashion a base will accept a proton according to:

$$B + H_3O^+ \overset{K_b}{\rightleftharpoons} BH^+ + H_2O$$

where B is the base, BH^+ is the conjugate acid, K_b is the dissociation constant of the base, and pK_b is the negative logarithm of the basic dissociation constant

$$pK_b = -\log_{10} K_b.$$

pK_a and pK_b are related by:

$$pK_a = 14 - pK_b.$$

This is derived from the equilibrium constant of water, discussed above.

pH Electrode

The primary standard for measurement of hydrogen ion concentration is the hydrogen electrode. This is a treated platinum electrode immersed in the solution to be measured and with gaseous hydrogen bubbled over its surface at an ambient pressure of 1 atmosphere. The e.m.f. of this electrode is zero when immersed in 1.0 M H_3O^+ ions. The e.m.f. at the electrode results from the ionization of hydrogen according to the equilibrium.

$$H_2 \rightleftharpoons 2H^+ + 2e^-.$$

The potential difference between this electrode and another standard electrode of known e.m.f. (e.g. a calomel electrode) is measured and used to calculate the H_3O^+ ion concentration.

The hydrogen electrode is not suitable for everyday use and has been generally replaced by the glass electrode.

Up to a few years ago it was usual to employ separate glass and reference electrodes to measure the pH of a solution. The glass electrode dipped directly into the solution, but the reference electrode had a ceramic plug which when immersed released a slow flow of KCl into the solution to form the salt bridge. Recently, combination glass and reference electrodes have become popular and are used for most routine pH measurements. The structural details of a combination electrode are shown in Fig. 7.4. The device is of concentric construction with the glass electrode being surrounded by the reference electrode. The virtue of this design is

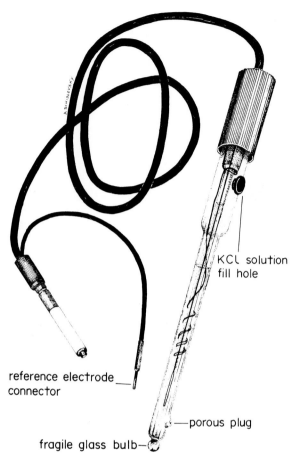

KCl solution
fill hole

reference electrode
connector

porous plug

fragile glass bulb

FIG. 7.4. The construction of a combined glass and
reference electrode used for pH measurement.

that besides being a compact and simple unit, the glass electrode is electrostatically screened over most of its length.

Care of the pH Electrode

(i) pH electrode bulbs are fragile and should be protected from physical damage.
(ii) Electrodes may be damaged by solutions of pH > 12 which tend to etch the glass.
(iii) Electrodes should be stored in water or dilute buffer solutions. If allowed to become dry, several hours of subsequent soaking in 1 N HCl will be needed before the electrode will function reproducibly.
(iv) Combined electrodes require the KCl bridge solution to be above the level of the solution being measured. This is to ensure an outward flow of KCl ions. The reference electrode compartment must be vented to the atmosphere.
(v) If fouled with protein or lipid solutions, electrodes may need cleaning with detergent or chloroform, these liquids can conveniently be applied with a cotton-wool swab. After this treatment the electrode should be soaked for 4–6 hours in 1 N HCl to restabilize the membrane hydration sheath.
(vi) When using a combined glass and reference electrode it is essential that the level of solution under test is high enough to cover the porous plug on the side of the electrode.

SECTION 8

FIXATION AND ASSIMILATiON OF NITROGEN FROM THE ATMOSPHERE AND COMBINED SOURCES

8.1. INTRODUCTION

Inorganic nitrogen is in general taken up by plants in the form of nitrate, although in some circumstances ammonium ions may be assimilated. The overall sequence of uptake of inorganic nitrogen into organic material can be resolved into the following sequence of reactions (Fig. 8.1).

(a) Nitrate reduction, via nitrite to ammonia.
(b) Ammonia assimilation into glutamate.
(c) Transamination from glutamate to primary amino acids.
(d) Synthesis of other amino acids.

In addition a number of free living bacteria and symbiotic micro-organisms are capable of

reducing atmospheric nitrogen gas to ammonia. In the case of legumes, and some other higher plants, the nitrogen fixing *Rhizobium* are contained in nodules on the roots. In all cases, in higher plants, energy is consumed in the reduction, synthesis and translocation of nitrogen from the roots to the growing regions, or storage organs. In the following sections the various techniques which may be used to determine the rates of nitrogen assimilation or to measure the activities of the enzymes involved are described in more detail.

8.2. NITROGEN FIXATION

by M. REPORTER and J. COOMBS

8.2.1. Introduction

Only certain prokaryotes, mainly bacteria and blue–green algae, can fix nitrogen. The organisms may be free living or symbiotic and include blue–green algae, free living bacteria and the symbiotic Rhizobia associated with legumes.

The basic metabolic requirements are:

1. A strong reducing agent of low redox potential (reduced ferredoxin, flavodoxin, NAD or NADP).
2. ATP, Mg^{2+}.
3. The enzyme complex nitrogenase.
4. Low or zero oxygen level.

The enzyme nitrogenase catalyses the reduction of N_2 to NH_3.

$$N_2 + 3XH_2 + 6ATP \rightarrow 2NH_3 + 3X_i + 6ADP + 6P_i$$

ASSIMILATION OF NITROGEN

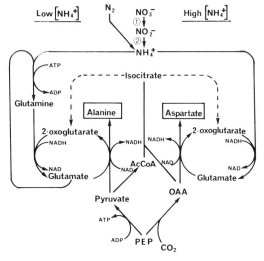

FIG. 8.1.

118

The enzyme has two major components, each of which has two or four sub-units. The larger component contains molybdenum, non-haem iron and sulphide, and the smaller contains iron and sulphide. Hence shortage of Mo or Fe can restrict N-fixing capacity. In addition nitrogenase is very easily inactivated by oxygen. The source of reductant varies greatly in different organisms, e.g. *Clostridium* supply electrons to ferredoxin through reduction of pyruvate. Rhizobia use aerobic oxidation of carbohydrate where as blue–green algae may use photosynthetically reduced ferredoxin, i.e.

nodules on the roots of members of the Leguminosae. It also includes bacteria which no longer possess an invasive characteristic, but which have an authentic history of origin from an invasive strain. In general, rhizobia only fix N_2 *in vivo* as bacteroids which contain the induced nitrogenase enzyme. However, some slow-growing strains may be cultured—these can fix N_2 *in vitro*.

Rhizobia are Gram-negative rod-shaped bacteria which often contain prominent granules of poly-B-hydroxybutyrate. The genus is divided into two classes: (a) the fast growers which

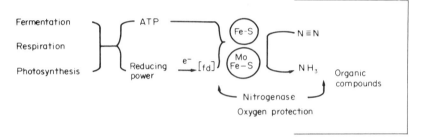

The activity of nitrogenase is measured either by following the assimilation of the heavy isotope ^{15}N, or more conveniently, using the acetylene reduction assay as detailed below.

8.2.2. The Genus *Rhizobium*

The genus *Rhizobium* contains those bacteria which are able to form morphologically distinct

have a mean generation time of 2 to 4 hour and form relatively large (2 to 4 mm diameter) colonies on agar media within 3 to 4 days and (b) slow growers which have a mean generation time of 6 to 8 hours and form small colonies, less than 1 mm in diameter, after 7 to 10 days. Table 8.1 lists various *Rhizobium* species and details the host genera which they preferentially invade.

Table 8.1. *Host preference of Rhizobium species*

Species	Preferred host genus
1. *Rhizobium leguminosarum*	*Pisum, Vicia, Lathyrus, Lens*
2. *R. trifolii*	*Trifolium*
3. *R. phaseoli*	*Phaseolus*
4. *R. meliloti*	*Medicago, Melilotus, Trigonella*
5. *R. lupini*	*Lupinus, Ornithopus*
6. *R. japonicum*	*Glycine max*
7. "Cowpea type"	*Vigna, Macroptilium* and others

Species 1–4 are fast growers and 5–6 are generally slow growers. The "cowpea rhizobia" group contains a diverse range of *Rhizobium* which cannot be accommodated in groups 1–6. They include both slow- and fast-growing strains, some of which may even infect non-leguminous angiosperms.

Detailed studies of the biochemical regulation of and/or limitations to carbon metabolism and nitrogen nutrition in nodules is essential if the symbiotic association is to be manipulated and improved. Room for improvement undoubtedly exists. For example, CO_2 enrichment of the foliar environment causes a significant increase in nitrogen fixation in nodulated soybeans, which strongly suggests that in ambient conditions the potential of nodules to fix nitrogen is limited by supply of photosynthate. The photosynthate supply is used not only as a source of carbon skeletons but also supplies the nodule with substrate to produce reductant and ATP for both the nitrogenase system and the synthesis of translocated nitrogen-containing compounds.

8.2.3. Assay of Nitrogenase Activity of Nodules from Leguminous Plants

Nitrogenase can be assayed by following the reduction of acetylene to ethylene because the nitrogenase enzyme complex reduces a number of substrates with triple bonds, the natural one being dinitrogen, i.e. $N \equiv N$. In an ideal situation rates of reduction of C_2H_2 should be confirmed using the $^{15}N_2$ heavy isotope method. If this is done it will be found that the moles of C_2H_2 reduced per mole of $^{15}NH_4$ formed may vary widely depending on conditions used. Usually the ratio of acetylene reduction to nitrogen reduction varies around the theoretical value of 3 to 5. However, higher values have been reported (Saito et al., 1979). These authors reported that when nodulated roots from different plants were used to determine the N_2 fixed and C_2H_2 reduction, on average the ratio $C_2H_2:N_2$ varied between 7.3 and 8.3. In vivo C_2H_4: $(3N_2 + H_2)$ electron balances were over one, varying from 1.32 to 1.43. These data suggested that overestimates of actual N_2 fixation might occur when using the C_2H_2 reduction technique, even if account was taken of H_2 evolution. Hence, care should be taken not to overestimate nitrogen fixation when using this method. On the other hand, the C_2H_2 reduction assay has the advantage that it requires cheaper, less complex apparatus, is quicker and easier to perform, and more widely adaptable to various experimental conditions, including assays in the field.

If possible it is a good idea to estimate H_2 evolution under the same conditions as used for C_2H_2 reduction in order to estimate the efficiency of the nodules being used. Nodules which are fixing nitrogen with a high efficiency evolve very little hydrogen. The relative efficiency (RE) of a given Rhizobium strain can be expressed as:

$$RE = 1 - \frac{H_2 \text{ evolved}}{C_2H_2 \text{ reduced}}.$$

(See Schubert and Evans, 1976.)

For detached nodules the assays may be carried out in calibrated serum vials fitted with rubber serum stoppers. A moist strip of filter paper (2×2 cm) is placed in each vial to prevent desiccation of the nodules. The nodules weighing between 100 mg and 1 g are placed on top of the paper and the vial sealed. In general the nodules should not be separated from the adjoining root piece. In order to obtain good rates of acetylene reduction the nodules should be solid, healthy looking and show red coloration (due to leghaemoglobin) if cut. For complete root systems larger glass vessels such as preserving (Kilner) jars fitted with a similar rubber serum stopper, inserted through the metal lid, may be used. In some cases the use of plastic bags may be suitable.

Acetylene is injected into each bottle to a concentration of 10% of the vial volume after withdrawing an equal volume of air. In the case of free living rhizobia the gas phase is replaced with a suitable mixture of argon, acetylene, carbon dioxide and oxygen in the ratio of $88.8:10:1:0.2$. In the case of anaerobic nitrogen fixers oxygen would not be included in this gas mixture and the assays would be conducted at about 1.1 atmospheres; i.e. with a positive pressure which prevents atmospheric oxygen entering the vessel.

The gas chromatographic assay for acetylene may be carried out using a Porapak N column, with helium as the carrier gas, and a flame ionization detector (see appendix). Column temperatures of 60° to 70°C are advisable with columns of 3 to 5 metres. Gas sampling is conveniently done using pressure-lock syringes (Precision Sampling Co.). The gas chromatographic assay for H_2 evolution is carried out in a similar manner but using a thermal conductivity detector and a molecular sieve 5A column approximately 6 metres long, with argon as carrier gas. The gas peak detected for C_2H_2 or H_2 on the gas chromatograph (GC) chart recorder is averaged from several determinations for each time point over a time course of up to 90 minutes, and compared with the peaks obtained using standard gas mixtures. With a rapid chart rate, measurement of the peak height may be sufficient to give a result of the required accuracy. Alternatively, if the recorder or GC is not fitted with automatic integration facility the areas may be determined by cutting out, from the chart paper, the area under the trace corresponding to a given gas peak. This piece of paper is then weighed and compared with the weight of a standard area (say 10 cm^2) of the same chart paper. The number of nmoles of H_2 or C_2H_2 in a given injection volume, at bottle pressure, is converted to a total number of nmoles in the bottle using the calibrated volume of each bottle or vial. The volume of any liquid in the vial is subtracted from the bottle volume. The nmole of gas per mg wet weight of the separated nodules (or per mg cell protein) is calculated per minute to give a standard rate for comparison.

8.2.4. Experimental

The Following Experiment Will Be Carried Out:

1. Record plant type, nodule pattern, number and size (weight).
2. Inspect one healthy-looking nodule and make sure that when cut it shows bright red leghaemoglobin. Old nodules are hollow and green; non-fixing nodules are white inside.
3. Take a vial and determine the volume (fill with water and weigh). The vial is then fitted with a tight serum stopper made of good rubber, else leaks of C_2H_2 and H_2 may cause problems.

4. If you have to detach the nodules do not cut from plant at the nodule base but cut off a small portion of root tissue with each nodule.
5. Place a moist strip of paper (2×2 cm) in the vial and place nodules on top.
6. Seal the vials and inject C_2H_2 prepared as described in the appendix (Burris, 1967). The amount of C_2H_2 added should be about 10% of vial volume. An equal volume of air should be removed through the stopper prior to addition of the C_2H_2. Note time of addition of acetylene.
7. Sample (0.25 cm^3) for C_2H_4 at least three time points (e.g. 15, 30 and 60 minutes) using a hypodermic syringe (gas tight) and inject into GC.
8. At end weigh the nodules.
9. Calculate the nmoles of acetylene reduced to acetylene and express as nmoles per g nodule (wet weight) per minute.

Bibliography and Further Reading

BURRIS, R. H. (1967) *In situ* studies on N_2 fixation using acetylene reduction technique. *Proc. Natl Acad. Sci. USA* **58**, 2071.

DILWORTH, M. J. (1966) Acetylene reduction by nitrogen fixing preparations from *Clostridium pasteuranium*. *Biochim. Biophys. Acta* **127**, 285.

HARDY, R. W. F., BURNS, R. C. and HOLSTEN, R. D. (1973) Application of acetylene-ethylene assay for measurement of nitrogen fixation. *Soil Biol. Biochem.* **5**, 47.

RAWSTHORNE, S., MINCHIN, F., SUMMERFIELD, R., COOKSON, E. C. and COOMBS, J. (1980) Carbon and nitrogen metabolism in legume root nodules. *Phytochemistry* **19**, 341.

REPORTER, M. (1978) Hydrogen evolution by rhizobia after synergetic culture with soybean cell suspensions. *Plant Physiology* **61**, 753.

SAITO, S. M. T., MATSUI, E. and SALATI, E. (1979) $^{15}N_2$ fixation, H_2 evolution and C_2H_2 reduction relationships in *Phaseolus*. *Physiol. Plant.* **49**, 37.

SCHUBERT, K. R. and EVANS, H. K. (1976) Hydrogen evolution—a major factor affecting the efficiency of nitrogen fixation in nodulated symbionts. *Proc. Natl Acad. Sci. USA* **73**, 1207.

Appendix A.8.1. Preparation of Acetylene

Acetylene may be generated from calcium carbide (CaC_2) using the following glass apparatus:

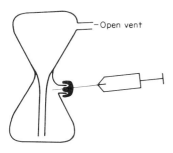

Procedure:

1. Completely fill lower resevoir with distilled water.
2. Drop in two CaC_2 pellets. To purge water of some dissolved air wait for a few seconds before stoppering with serum cap.
3. Withdraw portion of C_2H_2 using a syringe via serum cap.

N.B. Periodically clean vessel with HCl to remove $Ca(OH)_2$.

A.8.2. Gas Chromatography

The basic apparatus is as follows:

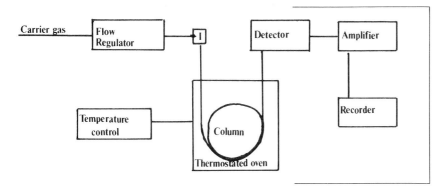

Carrier gas from a cylinder is passed through a flow regulator and past an injection port, I, where it picks up the sample for analysis. The injection should be as fast as possible—usually a special μ 1 syringe is used. The carrier gas and sample then pass through the column in a thermostatic oven where the components of the mixtures are separated. The columns, column packings and detectors used vary widely depending on the nature of the material being analysed. Discussion here will be restricted to the Poropak column packings and the flame ionization detector.

For a single substance a chromatogram might be obtained as follows:

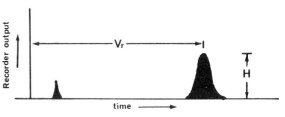

$$V_r = \text{retention volume,}$$
$$H = \text{peak height.}$$

The volume of carrier gas passed between the point of injection of the sample and the peak maximum, measured at the average pressure and temperature of the column, is termed the true retention volume of the substance (V_r). V_r is related to the retention time (t_r) (usually used to characterize a given substance) by the equation $t_r = V_r/F_{av}$ (where F_{av} = gas flow rate).

The retention volume of solvent, air or other component which moves with carrier gas equals the volume of gas in the column, denoted by V_g and may be called the dead volume of the column. The net retention volume V_r' is the true retention volume less the dead volume, i.e. $V_r' = V_r - V_g$.

Since the retention volume is measured at the temperature and pressure of the laboratory, the true retention volume has to be obtained from the experimental retention volume V_{exp} by the formula

$$V_r = V_{exp}(Tp_r)/(T_rP)$$

where T and P are the column and T_r and p_r the laboratory temperature and pressure respectively. The area of the peak depends on the amount of the substance present, the detector efficiency, and the degree of amplification used. If the latter factors are held constant the recorded peak area is a direct measure of the amount of substance present in the sample. Methods of determining the area vary from measurement by cutting out the peak from the paper and weighing it to electrical integration by microprocessor. This volume is then compared with that obtained using a standard sample.

Detector—Flame Ionization

In the flame ionization detector a hydrogen/nitrogen mixture is used as the carrier gas and is burnt in the detector at a jet above which a collector electrode is placed. The rate of ion production in the flame is measured by applying a potential of 100 to 300 V between the jet and the collector electrode. With only hydrogen burning in the flame the ionizing efficiency is very low and only about one in 10^{12} H_2 molecules produce an ion. Addition of an organic substance to the flame greatly increases the ionization. The ionization efficiency per carbon atom being about 1 in 10^5. The ions are collected at an electrode and the resulting current amplified. The minimum detectable concentration of organic material is about 1 part in 10^9 of hydrogen, and the detector has a linear response up to a concentration of about 0.5%. The linear range is about 10^7. The sensitivity and range makes it the ideal detector for use with organic materials.

Packing Material—"Porapak"

Porapak packing materials are porous polymer beads developed by the Dow Chemical Company, Texas, U.S.A. Eight different types are available identified in order of increasing polarity as types P, Ps, Q, etc., through to T. They are relatively robust but should not be used at temperatures above 250°C. With Porapak Q the water peak elutes between ethane and propane, hence hydrocarbon separations can be made where there is water contamination. This is, of course, important in the assay of nitrogenase activity using acetylene reduction.

The separation of various gases on "Porapak" materials is shown below.

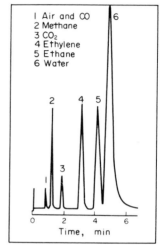

1 Air and CO
2 Methane
3 CO_2
4 Ethylene
5 Ethane
6 Water

Hydrocarbons in water

Column: 2 m × 2.3 mm; Support: Porapak Q, 80/100 mesh; Carrier: helium, 20 ml/min; Detector: TC; Temperatures: injector 150°C, column 50°C, detector 200°C.

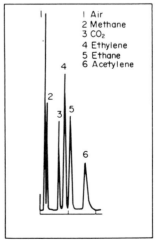

1 Air
2 Methane
3 CO_2
4 Ethylene
5 Ethane
6 Acetylene

$C_1 - C_2$ Hydrocarbons

Column: 1 m × 2.3 mm; Support: Porapak N, 80/100 mesh; Carrier: Helium, 25 ml/min; Detector: TC; Temperatures: Injector 150°C, column 50°C, detector 200°C.

8.3. ASSIMILATORY NITRATE REDUCTION

by M. G. GUERRERO

8.3.1. Introduction

For most plants in their natural environment, nitrate is the usual source of nitrogen. Nitrate must be converted to ammonia before it is combined with carbon compounds to form the various nitrogenous components of the cell (see Section 8.4). The process is known as assimilatory nitrate reduction to differentiate it from nitrate reduction of the respiratory type carried out by various bacteria which, under micro-aerophilic or anaerobic conditions, use nitrate as an electron acceptor in place of molecular oxygen. It can be estimated that plants assimilate 10^{10} tonnes of nitrate-N per year.

Assimilatory nitrate reduction takes place in higher plants, algae and many bacteria, yeasts and fungi. The process can be summarized as follows:

$$\overset{(+5)}{NO_3^-} \xrightarrow[\substack{\text{nitrate}\\\text{reductase}}]{+2e} \overset{(+3)}{NO_2^-} \xrightarrow[\substack{\text{nitrite}\\\text{reductase}}]{+6e} \overset{(-3)}{NH_4^+}$$

This involves the sequential participation of two metalloproteins—nitrate reductase and nitrite reductase. The physiological sources of electrons are reduced pyridine nucleotides or reduced ferredoxin according to the source and the type of enzyme. ATP is not required for nitrate or nitrite reduction. Both reactions take place with large decrease in free energy.

8.3.2. Nitrate Reductase

In eukaryotes, nitrate reductase is a multi-component enzyme complex (MW 200–300 thousand daltons) containing flavin (FAD), heme (cyt b_{557}) and molybdenum. Electrons provided by NAD(P)H (from photosynthesis or carbohydrate oxidation) are transferred to nitrate through the electron transport chain of the enzyme:

In addition to catalysing the reduction of nitrate by reduced pyridine nucleotides, the NAD(P)H:nitrate reductase exhibits two other activities which can be assayed separately and involve only part of the overall electron-transport capacity of the enzyme. Diaphorase activity results in the reduction by NAD(P)H of various 1- and 2-electron acceptors (cytochrome *c*, ferricyanide and other oxidants). The terminal nitrate reductase catalyses the NAD(P)H-independent reduction of nitrate by reduced flavin nucleotides or viologens. Both moieties participate jointly in the sequential transfer of electrons from NAD(P)H to nitrate:

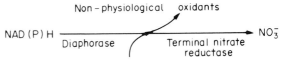

In eukaryotic algae and photosynthetic tissues of higher plants, nitrate reductase appears to be located either in the cytoplasm or loosely bound to the outer envelope of chloroplasts.

In prokaryotic photosynthesizing organisms, such as blue–green algae (cyanobacteria), nitrate reductase is a smaller protein (MW 75,000 daltons) containing molybdenum, but no flavin or heme. The enzyme appears tightly bound to thylakoids and uses photosynthetically reduced ferredoxin, but not NAD(P)H, as the electron donor. Cyanobacterial nitrate reductase can be envisaged as a precursor of the terminal moiety of the eukaryotic NAD(P)H:nitrate reductase.

Assimilatory nitrate reductase from different organisms, either prokaryotes or eukaryotes, has a fast rate of protein turnover, and is present at high level when cells are fed with nitrate, but repressed if the cells or plants are grown on media containing ammonium ions.

$H^+ + NAD(P)H$ ⎰ → $[FAD \longrightarrow cyt\ b \longrightarrow Mo] \longrightarrow$ ⎱ NO_3^-

$NAD(P)^+$ ⎱ $NO_2^- + H_2O$

8.3.3. Nitrite Reductase

The photosynthetic enzyme of both eukaryotic and prokaryotic cells is ferredoxin:nitrite reductase, which accepts electrons from photosynthetically reduced ferredoxin but not from reduced pyridine nucleotides.

The enzyme protein (MW about 60,000 daltons) contains a tetranuclear iron–sulphur centre and siroheme, a special type of heme similar to that present in sulphite reductase. The electron-transport chain from reduced ferredoxin to nitrite may be depicted as follows:

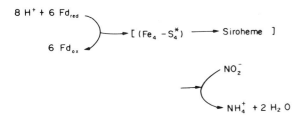

Nitrite is directly reduced to ammonia, without liberation of free intermediates. Photosynthetic nitrite reductase appears to be located within chloroplasts or thylakoids, it may also be detected in non-green tissue such as roots.

8.3.4. Photosynthetic Reduction of Nitrate

The process of assimilatory nitrate reduction in photosynthetic organisms may be summarized as shown in Fig. 8.2. In blue–green algae, nitrate reduction is intimately linked to photosynthesis, and its coupling with the photolysis of water can be conclusively demonstrated. The photoreduction of nitrate to ammonia with water as the electron donor can be considered as one of the simplest examples of photosynthesis. In eukaryotic green algae and higher plants, nitrite reduction is also a photosynthetic process and the reducing power for nitrate reduction, though not in such a direct way as for nitrite reductase, comes ultimately from the assimilatory power generated during the light reactions of photosynthesis.

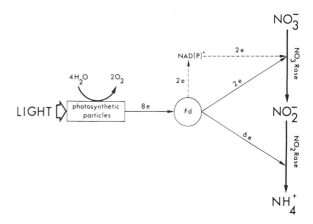

FIG. 8.2. The reduction of nitrate to ammonium by photosynthetically generated reductant.

8.3.5. Determination of Enzymatic Activities (Nitrate Reductase and Nitrite Reductase)

Nitrate reductase activity is usually estimated by measuring the amount of nitrite being produced from nitrate, and nitrite reductase activity by the disappearance of nitrite from the reaction mixture.

The method used for measuring nitrite (according to Snell and Snell, 1949) is based in the formation of a diazonium salt during the reaction, in acid medium, of nitrite with sulphanilamide, and the further establishment of a coloured complex as the diazonium salt reacts with N-(1-naphtyl)ethylenediamine (NNEDA). The resulting complex is pink in colour and has an absorption maximum at 540 nm (see diagram, page 126).

Nitrite estimation. Add 1 ml of 1% (w/v) sulphanilamide in 2.4 N HCl to 3 ml of the sample containing nitrite (10 to 100 nmole). Add 1 ml of 0.02% (w/v) N-(1-naphtyl)ethylenediamine dichloride to each tube. Mix (Vortex mixer) and measure absorbance at 540 nm after 10 minutes.

$$E_{1\,cm}^{1\,mM} \text{ nitrite complex (540 nm)} = 55,$$

where E is the extinction at the given wavelength.

8.3.5.1. Nitrate Reductase Activity
A. "*In vivo*" Assay of Nitrate Reductase in Plant Leaf Tissue

See Scott and Neyra (1979), Klepper *et al.* (1971).

COLORIMETRIC METHOD FOR NITRITE DETERMINATION

COLOURED COMPLEX (λ_{max} at 540 nm)

($E_{1cm}^{1mM} = 55$)

Cut the leaves into pieces (squares of about 5 mm²). Add 0.2 g to 5 ml of incubation mixture containing: 0.1 M potassium phosphate buffer, pH 7.7, 0.1 M KNO₃, and 1% (v/v) isopropanol. Flush the mixture with argon for about 5 minutes. Incubate in the dark at 30°C. Remove aliquots with hypodermic syringe (0.2–1 ml) for nitrite determination at zero time and after 15–30 minutes.

This method can also be used for estimating the nitrate content of leaves, following the same procedure but omitting nitrate from the reaction mixture. Allow the reaction to proceed until reaching a constant concentration of nitrite (1 to 2 hours).

B. "*In situ*" Assay of Nitrate Reductase Activity in Algal Cells

Green Algae. For *Chlorella* use the freezing–thawing procedure described by Hipkin and Syrett (1973) and by Syrett and Thomas (1973), as follows:

(a) Collect cells by centrifugation (500 × g for 2 minutes).
(b) Wash the cells with 0.067 M phosphate buffer, pH 6.8.
(c) Resuspend the cells in 0.1 M HCl-Tris buffer, pH 8.5 to reach a concentration of about 2 mg dry wt. cells per ml buffer.

(d) Freeze overnight (about 16 hours at −20°C).
(e) Thaw at about 3°C, centrifuge and resuspend in a total volume of 1.5 ml of the reaction mixture which contains:

HCl-Tris buffer, pH 8.5	100 μmol
KNO₃	10 μmol
Malate	75 μmol
NADH	0.03 mg

Note: NADH is not required when the species of *Chlorella* is *C. fusca*.

(f) Incubate for 10 minutes at 30°C.
(g) Add 0.25 ml of 25% (w/v) ZnSO₄ and 0.25 ml of 1 M NaOH. Let stand for 5 minutes at 0°C, and centrifuge.
(h) Use an aliquot (1 ml of the supernatant) for nitrite estimation as indicated above.

For assays with cells of other strains of green algae (*Ankistrodesmus braunii* and *Chlamydomonas reinhardii*) see Hipkin and Syrett (1977).

Blue–green algae (Cyanobacteria): Works very well with *Anacystis, Anabaena* and *Nostoc*:

(a) Add 20 μl toluene to 1 ml of cell suspension (either in buffer or in culture medium) containing about 10 to 20 μl packed cell volume cells.

(b) Shake vigorously (Vortex mixer) for 90 seconds.

(c) Add a 0.1 ml aliquot of the suspension to 0.8 ml of standard assay mixture for ferredoxin–nitrate reductase (MV˙ as reductant) activity determination:

0.5 M HCO_3^-/CO_3^- buffer, pH 10.5	6.25 ml
KNO_3	63.1 ml
methyl viologen (MV)	32.2 mg
water up to	25 ml

(d) Mix gently and add 0.1 ml of freshly prepared dithionite solution containing 20 mg $Na_2S_2O_4$ per ml of 0.3 M $NaHCO_3$.

(e) Mix gently and incubate for 5 minutes at 30°C.

(f) Stop the reaction by shaking in Vortex mixer until blue colour (reduced methyl viologen) disappears.

(g) Add 0.1 ml of 25% $ZnSO_4$ and 0.1 ml 1 M NaOH.

(h) Centrifuge (bench centrifuge) and determine nitrite as described above using an aliquot (0.6 ml) of the supernatant.

C. "In vitro" Assays of Nitrate Reductase Activity

Follow the general recommendations given in Section 9.3 for assay and extraction of enzymes. For nitrate reductase the presence of EDTA and dithioeythritol in the extraction buffer is recommended, as well as FAD (about 20 μM) in the case of NAD(P)H-dependent enzymes.

Ferredoxin–Nitrate reductase. From blue–green algae.

Use reduced methyl viologen as reductant as described by Manzano et al. (1976) with slight modifications. Use the same reaction mixture and assay conditions as described above for the in situ assay (omit toluene treatment).

NAD(P)H–nitrate reductase. For the enzyme from green algae, higher plants and fungi.

(a) Mix

0.5 M HCl–Tris buffer, pH 7.5	0.2 ml
0.2 M KNO_3	0.1 ml
3 mM NADH or NADPH	0.1 ml
enzyme preparation + water	0.3 ml

(b) Start the reaction by addition of the enzyme preparation.

(c) Incubate at 30°C for 5 to 15 minutes.

(d) Stop the reaction by addition of 0.1 ml 10 μM phenazine methosulphate to oxidize the remaining NAD(P)H. After 5 minutes add 0.1 ml 25% $ZnSO_4$ and 0.1 ml 1 M NaOH. Centrifuge and remove 0.65 ml of the supernatant for nitrite estimation.

In many cases the presence of FAD in the reaction mixture (20 μM final concentration) is required for achieving maximal activity.

The reaction can also be assayed spectrophotometrically by following the oxidation of NAD(P)H at 340 nm. Use 3-ml cuvettes containing 0.6 ml of the Tris buffer + 0.3 ml of 1 mM NADH (or NADPH) + 1.7 ml water. Start the reaction by adding 0.3 ml of enzyme preparation. Follow the decrease in absorbance at 340 nm

$$E_{1\,cm}^{1\,mM}\ NAD(P)H\ (340\ nm) = 6.3.$$

8.3.5.2. Nitrite Reductase Activity ("in vitro" assay)

Ferredoxin–nitrite reductase. For the enzyme from algae and higher plants.

Measure disappearance of nitrite using reduced methyl viologen as electron donor.

Reaction mixture:

0.5 M HCl–Tris buffer, pH 7.5	6.25 ml
$NaNO_2$	4.32 mg
Methyl viologen	6.01 mg
Water up to	25 ml

Procedure. Add 0.3 ml of the enzyme preparation to 1.5 ml of the reaction mixture. Run a blank without enzyme. Start the reaction by adding 0.2 ml of a recently prepared dithionite solution containing 25 mg $Na_2S_2O_4$ per ml of a $NaHCO_3$ solution containing 25 mg bicarbonate per ml. Incubate for 15 minutes at 30°C. Stop the reaction by vigorous shaking (Vortex mixer) until blue colour disappears. Use a 20-μl aliquot for nitrite determination. Estimate the amount of nitrite which has disappeared using as reference the assay without enzyme.

NAD(P)H–nitrite reductase. For the enzyme from fungi and some bacteria.

Use a similar reaction mixture to that described for NAD(P)H–nitrate reductase activity assay, but containing $NaNO_2$ instead of nitrate. Follow the nitrite-dependent decrease in absorbance at 340 nm. FAD (about 20 μM) is usually required for maximal activity.

8.3.6. Other Analytical Procedures Useful for the Study of Nitrate Assimilation

Sometimes it is interesting to follow the disappearance of nitrate or the appearance of ammonium in a cell suspension or in a sample of soil, or even to determine the concentration of these compounds in plant tissues or algal cells. Chemical methods are described for the determination of nitrate and ammonium. Ammonium can also be determined enzymatically with glutamate dehydrogenase, and nitrate with nitrate reductase. In both cases the oxidation of NAD(P)H which is dependent upon the presence of the respective substrates can be followed spectrophotometrically by measuring the decrease in absorbance at 334 or 340 nm. A method for the enzymatic determination of ammonium will also be described.

8.3.6.1. Determination of Nitrate

After Cawse (1967). The method is based in measurement of the ultraviolet (210 nm) absorption due to nitrate, after the removal of interference due to other ions (mainly nitrite) by treatment of the sample with perchloric and sulphamic (amidosulphuric) acids.

Add 0.1 ml of 10% (w/v) sulphamic acid to 1.5 ml of the sample containing nitrate (10 to 200 nmole). Shake (Vortex mixer), let stand for 2 minutes, shake again and add 0.4 ml of 20% (v/v) perchloric acid. Shake again, and measure the absorbance at 210 nm using silica cells

$$E_{1\,cm}^{1\,mM}\ \text{Nitrate (210 nm)} = 7.4$$

8.3.6.2. Chemical Determination of Ammonium and Dissolved Ammonia

After Solorzano (1969). The method is based on the formation of indophenol after the reaction, at high pH, of ammonia, phenol and hypochlorite. Possible interference due to calcium or magnesium are eliminated by treatment with citrate. Under the final assay conditions, indophenol exhibits a blue colour with an absorbance maximum at 640 nm.

Solutions:

(1) Phenol–alcohol solution: 0.1% (w/v) phenol in 95% (v/v) ethanol.
(2) Nitroprusside solution: 0.5% (w/v) sodium nitroprusside (store in amber bottle for not longer than 1 month).
(3) Alkaline solution: 20% (w/v) trisodium citrate in 1% (w/v) NaOH solution.
(4) Oxidizing solution: 16 ml of solution (3) plus 0.4 ml of commercial hypochlorite solution (1.5 N at least) plus 3.6 ml water. This oxidizing solution should be prepared just before using it (use it within the same day).

Procedure:

To 1.6 ml of the sample containing ammonia (up to 80 nmol) add 0.2 ml of solution (1), then 0.2 ml of solution (2), and finally 0.5 ml of solution (4), mixing thoroughly after each addition. The colour is allowed to develop for 60 to 90 minutes at room temperature in the dark. Finally, absorbance at 640 nm is measured.

$$E_{1\,cm}^{1\,mM}\ \text{Indophenol–nitroprusside (640 nm)} = 18.5.$$

Precautions:

Newly distilled water must be used, and the glassware be cleaned by washing with warm diluted HCl and rinsing thoroughly with distilled water.

In either nitrate or ammonium determinations, blanks without the corresponding N-compound should be run in parallel. Run also standards containing about 50 nmol of the corresponding N-compound.

8.3.6.3. Enzymatic Determination of Ammonium and Dissolved Ammonia

With glutamate dehydrogenase, by estimating the oxidation of reduced pyridine nucleotide. See Section 8.4.4.

Solutions:

(1) 0.15 M Triethanolamine buffer, pH 8.6.
(2) 0.3 M α-ketoglurate (2-oxoglutarate).
(3) 30 mM ADP.
(4) 2.4 mM NADPH.

Mix 1.45 ml of solution (1), 0.1 ml of solution (2), 0.1 ml of solution (3) and 0.2 ml of solution (4). Add up to 1.1 ml of the sample containing ammonium (up to 300 nmol) and complete the volume to 2.95 ml with water. Measure absorbance at 340 nm (should be about 1.0). Add 50 μl of an ammonium-free preparation of glutamate dehydrogenase containing about 400 units per ml. Wait 10 minutes. Measure A_{340} again. Repeat the measurement every 5 minutes till the value of absorbance does not change more than 0.01–0.02.

A blank without ammonium should be run in parallel, and also a standard containing 100 nmol ammonium.

$$E_{1\,cm}^{1\,mM} \text{ NADPH (340 nm)} = 6.3,$$

$$\frac{(A_{340} \text{ sample} - A_{340} \text{ blank}) \times 3}{6.3} = \mu\text{mol ammonium}$$

in the assay.

8.3.7. Uptake of Nitrate by Cells of Blue–Green Algae. Effects of Ammonium and Methionine Sulphoximine

Purpose: (1) Follow the utilization of nitrate by cells of *Anacystis nidulans* estimating the disappearance of the ion from the medium. (2) Demonstrate the short-term inhibitory effect of ammonium on nitrate uptake. (3) Show that the ammonium effect requires ammonium to be metabolized, and that the inhibition of nitrate uptake does not take place in cells treated with methionine sulphoximine, an inhibitor of glutamine synthetase (see Section 8.4.5).

Experimental

Assays of nitrate uptake will be run in air-opened conical flasks with continuous shaking in the light at 40°C, containing cell suspensions of *Anacystis* in 25 mM Tricine–NaOH buffer, pH 8.3 (about 2 μl packed cell volume per ml buffer). Methionine sulphoximine will be added to one of the flasks to reach a final concentration of 1 mM. After 15 minutes of preincubation the assay will be started by the addition of nitrate (0.4 mM final concentration). Two blanks (plus and minus methionine sulphoximine without added nitrate will be run in parallel. At times zero, 15 minutes and 30 minutes, 10-ml aliquots will be withdrawn. At time 30 minutes NH$_4$Cl will be added to each flask to reach 0.1 mM final concentration. Samples will be withdrawn at $t = 30$ minutes, 45 minutes and 60 minutes.

The aliquots taken at different times will be centrifuged immediately after sampling to remove the cells. The corresponding supernatants will be stored in a refrigerator until the determinations of nitrate and ammonium are carried out.

0.4-ml aliquots will be used for estimation of nitrate and 0.6-ml aliquots for ammonium determination.

The obtained values for the concentration of nitrate and ammonium in medium will be plotted against time.

Bibliography and Further Reading

CAWSE, P. A. (1967) The determination of nitrate in soil solutions by ultraviolet spectrophotometry. *Analyst* 92, 311.

HIPKIN, C. R. and SYRETT, P. J. (1973) Enzymic determination of nitrate by use of frozen thawed *Chlorella* cells. *New Phytol.* 72, 47.

HIPKIN, C. R. and SYRETT, P. J. (1977) Nitrate reduction by whole cells of *Ankistrodesmus braunii* and *Chlamydomonas reinhardi*. *New Phytol.* 79, 639.

KLEPPER, L., FLESHER, D. and HAGEMAN, R. H. (1971) Generation of reduced nicotinamide adenine dinucleotide for nitrate reduction in green leaves. *Plant Physiol.* 48, 580.

LOSADA, M. and GUERRERO, M. G. (1979) *Topics in Photosynthesis*, Vol. 3. (Ed. J. BARBER), pp. 365. (Elsevier: Amsterdam.)

MANZANO, C., CAUDAU, P., GOMEZ MORENO, C., RELINIPO, A. M. and LOSADA, M. (1976) Ferredoxin dependent photosynthetic reduction of nitrate and nitrite by particles of *Anacystis nidulans. Molec. Cell. Biochem.* **10**, 161.

SCOTT, D. B. and NEYRA, C. A. (1979) Glutamine synthetase and nitrate assimilation in sorghum (*Sorghum vulgare*) leaves *Can. J. Bot.* **57**, 754.

SNELL, F. D. and SNELL, C. T. (1949) *Colorimetric Methods of Analysis*. (Van Nostrand: N.Y., London.)

SOLORZANO, L. (1969) Determination of ammonia in natural waters by the phenol hypochlorite method. *Limnol. Oceanogr.* **14**, 799.

SYRETT, P. J. and THOMAS, E. M. (1973) The assay of nitrate reductase in whole cells of *Chlorella*, strain differences and the effect of cell walls. *New Phytol.* **72**, 1307.

VENNESLAND, B., GUERRERO, M. G. (1979) *Encyclopedia of Plant Physiology*, New Series, Vol. 6. (Eds. M. GIBBS and E. LATZKO), pp. 425. (Springer: Berlin.)

8.4. THE ASSIMILATION OF AMMONIA

by P. J. LEA

8.4.1. Introduction

Ammonia is synthesized by the reduction of nitrogen gas or nitrate as described in the previous sections. Ammonia is also synthesized in a number of other reactions particularly in the conversion of glycine to serine in photorespiration (see Section 4.4) and following the breakdown of protein when glutamate and arginine are deaminated. After the transport of nitrogen to the developing seed ammonia is again liberated on the breakdown of asparagine and the ureides. Until 1974 it was assumed in plants, that ammonia was assimilated via the enzyme glutamate dehydrogenase:

$$NH_3 + 2\text{-oxoglutarate} + NAD(P)H_2$$
$$\rightleftharpoons glutamate + H_2O + NADP$$

The enzyme is particularly abundant in plants and can be assayed in all tissues apart from certain blue–green algae. Current evidence suggests that the enzyme is not involved in ammonia assimilation, but that it may be involved in the catabolism of amino acids to yield ammonia and readily oxidizable 2-oxo-acids, particularly in the germinating seed. However, the

precise role of the enzyme is not clear at the present time.

Data obtained from a wide range of workers and using a variety of techniques now confirm that under normal growth conditions ammonia is assimilated by a two-step reaction involving the action of two enzymes:

(1) Glutamine synthetase
$$NH_3 + glutamate + ATP$$
$$\rightarrow glutamine + H_2O + ADP + P_i.$$

(2) Glutamate synthase (GOGAT)
glutamine + 2-oxoglutarate + NAD(P)H$_2$ or reduced ferredoxin → 2 molecules of glutamate + NAD(P) or oxidized ferredoxin.

Glutamine synthetase is localized in the cytoplasm and the chloroplast of leaves but GOGAT is situated only in the chloroplast. When ammonia is assimilated in the chloroplast the reaction is powered by ATP and reduced ferredoxin generated by the light reactions. In this case ammonia assimilation and nitrite reduction can be termed PHOTOSYNTHESIS in a similar manner to CO_2 fixation. In roots and maturing seed tissue the GOGAT enzyme can utilize either reduced ferredoxin or reduced coenzyme. The precise electron donor *in vivo* has not yet been ascertained.

8.4.2. Amino Acid Synthesis

The amino group of glutamate may be readily transferred to form aspartate and alanine from oxaloacetate and pyruvate respectively, e.g.

glutamate + oxaloacetate
$$\rightleftharpoons 2\text{-oxoglutarate} + aspartate$$

The reactions are carried out by amino acid transferases, and all the other protein amino acids can be formed from glutamate, aspartate and alanine provided the correct 2-oxo acid is available. In this manner 2-oxoglutarate is regenerated to take part in the GOGAT reaction. The three major 2-oxo acids are synthesized in the mitochondria in the tricarboxylic acid cycle but the majority of the other carbon skeletons can be synthesized in the chloroplasts in photosynthetic reactions.

8.4.3. Enzyme Assays

Unless absolutely essential it is not recommended that any assays are carried out on totally crude extracts. Plant extracts contain a large number of inhibitors (in particular oxidizable phenols) which will result in an underestimate (or even a zero determination) of the enzyme activity being measured. It must be remembered that just because an enzyme cannot be detected in an extract under one set of conditions, it does not mean that the enzyme is not present. It should also be noted that a set of conditions that is optimum for one enzyme is not necessarily correct for another.

Extraction

A number of different buffer solutions should be tested. Suggested starting points are Tris/HCl, Tricine/KOH, Imidazole/KOH or K phosphate at 50–100 mM. The inclusion of glycerol (10–20%) and a sulphydryl reagent 2-mercaptoethanol (10–20 mM) is recommended. EDTA (1–5 mM) may be used to chelate metal ions but care must be taken to ensure that any ions that activate the enzyme under test are replaced. If the plant material is known to have a high proteolytic activity (which may breakdown the enzyme during the isolation procedure) the inclusion of 2% casein (particularly important for nitrate reductase) or an inhibitor phenylmethyl sulphonyl fluoride (1 mM dissolved in isopropanol initially) is recommended. If the plant material is known to have a high phenol content and extracts can be seen to go brown during the extraction procedure then insoluble polyvinyl pyrrolidone (5%) or sodium diethyl dithiocarbamate may be included. If the enzyme is known to be activated by a metal ion (e.g. K^+ or Mg^{++}) then it should be included at all times in the extraction medium. It can be seen that there are a number of compounds that may be added to the extraction medium and it is very important that where a new enzyme or new tissue is being investigated, that time is spent making sure that the correct isolation procedure is employed and that the maximum amount of enzyme is being assayed. In the following sections a typical isolation buffer will be suggested, but this is only a guide, and there may be considerable variations between different tissues as well as different plants.

All extractions should be carried out below 4°C, and depending on the tissue may be ground (by hand) in a pestle and mortar or homogenized in a kitchen blender. A particular useful (but very expensive) apparatus named a "Polytron" utilizes the principles of sonication and high-speed mechanical homogenization at the same time. Fibrous tissue should preferably be cut into small pieces before extraction and the inclusion of some acid-washed sand may help to extract the maximum amount of enzyme. If fibrous tissue is first cooled in liquid nitrogen the tissue becomes very brittle and may easily be ground into very small pieces prior to extraction. Dry seeds should, of course, be ground to a fine flour before extraction. With fresh tissue it is very important that extraction takes place as soon as possible after harvesting of material. This is particularly important with root and root nodule tissue which can frequently be seen to go brown in a very short time. Plant material may be stored deep frozen (−20°C), but as enzyme activity can decrease with time, the use of liquid nitrogen (−196°C) for storage is recommended.

The buffer volume used will again vary with tissue but 10 ml per gramme of tissue is probably the maximum used. After complete homogenization the extract should be squeezed through two to four layers of muslin (cheesecloth, "Miracloth") and centrifuged at 10,000 g for 15 minutes. The supernatant should be devoid of all unbroken cells and cell wall tissue. There will, however, be a certain amount of membrane tissue from broken organelles. The extract should contain all the soluble enzymes in the cytoplasm, vacuole, chloroplast mitochondria, etc.; however, enzymes known to be attached to membranes may have to be removed by detergents. The precipitate should be tested for enzyme activity (if possible) to ensure that there is no large amount of enzyme present in the pellet.

The supernatant should be made 65% saturated with ammonium sulphate (see Section 9.8) and the precipitated protein collected by centrifugation. The resuspended pellet should be passed through a column of Sephadex G.25 (see Section 9.9) equilibrated in extraction buffer. If the enzyme is known to have a molecular weight of more than 100,000 then Sephadex G.75 is probably a better resin to use. Columns of 1.5 cm diameter by 20 cm length will easily accommodate 3 ml of resuspended ammonium sulphate pellet. The first protein containing fractions eluting from the column should be collected and used for the enzyme assays.

8.4.4. Glutamate Dehydrogenase

Extraction buffer. Tris/acetate (50 mM, pH 8.2), 10% glycerol, 0.5 mM EDTA, 5mM mercaptoethanol. The assay involves measuring in a recording spectrophotometer the decrease in absorption at 340 nm (the wavelength at which NAD(P)H absorbs) following the oxidation of NAD(P)H. The reaction mixture contains in a final volume of 2.5 ml: 370 μmol of ammonium acetate, 200 nmol NADH, 31 μmol 2-oxoglutarate, 2 μmol CaCl$_2$ and 125 μmol Tris/acetate buffer pH 8.2. All solutions are adjusted to pH 8.2 with Tris. Activity is determined by measuring the difference between the rate of NADH oxidation in the presence and absence of ammonium acetate. The actual rate of enzyme activity can be determined; it is known that a solution of NADH of 1 μmol/ml has an absorbance of 6.2. Two important points to note are:
(1) The concentration of ammonium ion must be high to ensure that the enzyme is fully saturated;
(2) There must be excess of a divalent cation, to activate the enzyme completely. It is not possible to make an accurate determination of glutamate dehydrogenase activity in the presence of EDTA alone, as the majority of the enzyme activity would be inhibited.

8.4.5. Glutamine Synthetase

Extraction buffer. Imidazole-HCl (50 mM, pH 7.2), 0.5 mM EDTA and 1 mM dithiothreitol.

Glutamine synthetase is frequently assayed in the presence of glutamine, hydroxylamine, ADP, arsenate and Mn^{2+}. This so-called "transferase" assay is NOT RECOMMENDED as the physiological relevance to glutamine synthetase is not known. The assay is, however, widely used mainly because very high activities may be determined.

A more physiological but by no means perfect assay is the "synthetase" assay which employs hydroxylamine in the place of ammonia.

glutamate + hydroxylamine + ATP
$$\rightarrow \text{glutamyl–hydroxamate} + ADP + P_i$$

The assay consists of incubating 0.25 ml of enzyme in a final volume of 1 ml containing 18 μmol ATP, 45 μmol MgSO$_4$, 6 μmol hydroxylamine, 92 μmol L-glutamate and 50 μmol Imidazole–HCl at a final pH of 7.2 for varying times at 30°C. The reaction is stopped by the addition of 1 ml of ferric chloride reagent (0.37 M ferric chloride, 0.67 M HCl and 0.2 M TCA) which forms a brown-coloured complex with any glutamyl–hydroxamate formed and precipitates out the enzyme protein. The reaction tubes are centrifuged to remove the protein, and the absorbance may be read in a spectrophotometer at 540 nm. A standard curve of glutamyl-hydroxamate may be prepared with up to 3 μmol per 1 ml of assay mixture. The rate of enzyme activity may then be calculated. For very active enzyme preparations it may be necessary to dilute the enzyme or use very short assay times as the production of the hydroxamate is not linear at very high rates. It is not known in crude extracts how close an agreement there is between the true rate of glutamine synthesis and the production of the hydroxamate complex. Two other assays may be used in purer extracts: (1) the determination of the phosphate liberated from ATP, and (2) the determination of ADP based on a spectrophotometric assay involving pyruvate kinase and lactate dehydrogenase.

Inhibitors of Glutamine Synthesis
A frequently used piece of evidence for the operation of the glutamine synthetase/glutamate

synthase pathway is that ammonia assimilation in plants is blocked by L-methionine sulphoximine (MSO).

$$CH_3$$
$$|$$
$$O = S = NH$$
$$|$$
$$CH_2$$
$$|$$
$$CH_2$$
$$|$$
$$CHNH_2$$
$$|$$
$$COOH$$

The compound acts as an inhibitor by binding to the active site of glutamine synthetase in place of the activated intermediate γ-glutamyl phosphate (Ronzio *et al.*, 1969). When MSO is fed to plants ammonia is rapidly evolved, and there is little evidence to show that glutamate dehydrogenase is able to assimilate the available ammonia. At low concentrations of MSO the inhibition is competitive with respect to glutamate.

Figures 8.3 and 8.4 show data obtained by Mr Mark Leason at Rothamsted working with glutamine synthetase isolated from pea leaves. In Fig. 8.3 the data are presented as the classical Lineweaver–Burke plot at various concentrations of MSO, of the reciprocal of the enzyme activity against the reciprocal of glutamate concentration. The V_{max} of the enzyme at an infinite glutamate concentration is the intercept on the

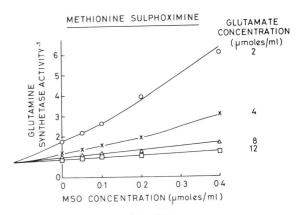

FIG. 8.4.

y-axis with a value of approximately 0.8. This value is independent of the amount of inhibitor present indicating that it is a COMPETITIVE type of inhibition. If the V_{max} had been lower with increasing inhibitor concentration then the inhibition would be considered NON-COMPETITIVE. At low concentrations of MSO the plots are linear, but at higher levels an upward curve can be detected, suggesting that there is irreversible binding of MSO to the active site. The slope of the non-inhibited line is equal to

$$\frac{K_m}{V_{max}}$$

but in the presence of inhibitor the slope is altered to

$$\frac{K_m}{V_{max}}\left(1 + \frac{[I]}{K_i}\right)$$

where $[I]$ is the concentration of the inhibitor and K_i is the inhibitor constant. By measuring the slopes of the lines it is possible to calculate the K_i value.

A more simple method of plotting the data is in the Dixon plot as shown in Fig. 8.4, of the reciprocal of enzyme activity against inhibitor concentration. The lines should all cross at a point that is vertically above the negative K_i value of the inhibitor. Thus in Fig. 8.4 the K_i can be read off directly as approximately 0.12 μmol/ml. Compounds can be compared for their potency as inhibitors by comparison of their K_i values, the lower the value the stronger the inhibitor.

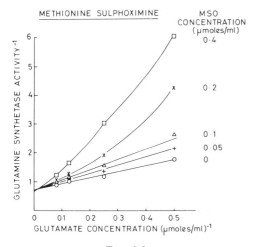

FIG. 8.3.

8.4.6. Glutamate Synthase

Extraction buffer. K phosphate (50 mM, pH 7.2), 5 mM EDTA, 12.5 mM 2-mercaptoethanol, 1 mM phenylmethyl sulphonyl chloride and 10% glycerol.

Pyridine Nucleotide-dependent Enzyme

The enzyme in roots, root nodules and developing fruits may be assayed by measuring the decrease in absorbance of NAD(P)H at 340 nm in a similar manner to glutamate dehydrogenase.

The assay medium should contain 12.5 μmol glutamine, 12.5 μmol 2-oxoglutarate, 200 nmol NADH and 125 μmol Tricine–KOH pH 7.5 in 2.5 ml. The enzyme activity is determined in the presence and absence of glutamine. Divalent cations are not required for the reaction and it is normal to have EDTA present in the reaction medium, to remove any other ions.

Ferredoxin-dependent Enzyme

The enzyme in leaf tissue is only ferredoxin-dependent but the enzyme from other sources is apparently able to use either coenzyme or reduced ferredoxin. It has not yet been possible to ascertain whether one or two enzymes are responsible for these two reactions.

The assay mixture contains 10 mM 2-oxoglutarate, 10 mM glutamine, 100 μg of methyl viologen and up to 100 μl of enzyme in a final volume of 0.5 ml. For assaying crude extracts 10 mM aminooxyacetate may be added to inhibit any transaminase activity. The reaction is started by the addition of 100 μl of a solution containing 16 mg sodium dithionite and 16 mg sodium bicarbonate per ml (prepared immediately before use). After incubation at 30°C for 5–15 minutes the reaction is stopped by the addition of 1 ml of ethanol followed by vigorous shaking to oxidize the methyl viologen and dithionite. A minus dithionite blank is always run for each reaction. Methyl viologen is used as a substitute for ferredoxin as it is much cheaper, and is very stable on storage. It also has the added advantage that it forms a blue colour when reduced, and its oxidation to a colourless form may be followed in the assay

tube. Ferredoxin may be used in place of methyl viologen in the assay mixtures and reaction rates are in general faster although care has to be taken to ensure that the ferredoxin is in a fully reduced state. After centrifugation 200 μl of the reaction mixture is spotted on to Whatman No. 4 chromatography paper. The separation of glutamate may be carried out by chromatography in 75% (w/w) phenol in the presence of ammonia vapour. The papers should be dried thoroughly after use and sprayed with a freshly prepared specific ninhydrin reagent suggested by Atfield and Morris (1961). The reagent comprises 0.05 g cadmium acetate, 1.0 ml acetic acid and 5.0 ml of H_2O in 50 ml acetone containing 0.5 g ninhydrin. The papers should be left in the dark in an enclosed container for a period of 12–18 hours in the presence of concentrated sulphuric acid. The glutamate spots may then be cut out and the dark-red colour eluted with 8 ml of a reagent containing 600 ml ethyl acetate, 600 ml water, 600 ml methanol, 18 g glacial acetic acid and 18 g cadmium acetate. The absorbance of the coloured solutions may be determined at 500 nm. A standard curve of glutamate concentrations should be made, and an internal standard of 20 μl of a 1 mg/ml solution should always be run on EACH chromatography paper, to allow for variations in relative humidity and temperature during development. There is a linear relationship between the absorbance at 500 nm and the glutamate applied over a range 0–50 μg.

Glutamate may also be separated from glutamine using Dowex-1 columns. Dowex-1-chloride form may be converted to the acetate form by thorough washing with 10% Na_2CO_3, 2 M acetic acid followed by excess distilled H_2O. Small columns (0.5 × 5.0 cm) of Dowex-1-acetate may be prepared in glass pasteur pipettes plugged at the tapered end with glass wool; 0.5 ml of the assay medium may be loaded on the columns and glutamine is eluted with 4 ml of distilled H_2O. Glutamate may then be eluted with 4 ml of 0.2 M acetic acid. The glutamate content of the eluate may be determined by any ninhydrin method provided that a standard glutamate curve is first constructed.

8.4.7. Amino Transferases

For an excellent review on this subject the reader is referred to Wightman and Forest (1978).

Extraction buffer. Tris/HCl (pH 7.5, 40 mм), EDTA 0.25 mм, glutathione 2 mм.

All aminotransferases are reversible, therefore the direction chosen may depend upon the substrates available. The enzymes are pyridoxal phosphate requiring, although it is now believed that in plants the coenzyme is bound tightly to the enzyme. The requirement for pyridoxal phosphate for the enzyme under test should be checked. Probably the simplest method of testing for aminotransferase is to incubate the enzyme with 5 mм 2-oxo acid and 5 mм amino acid in 50 mм Tris/HCl buffer pH 7.5 for varying times and stopping the reaction with an equal volume of ethanol. After centrifuging the protein, the extract may be chromatographed on paper or TLC plates in a solvent that gives a good separation of the initial end product amino acid (e.g. butanol:acetic acid:water 90:10:29 by volume gives a good separation of amino acid aspartate and alanine). The rate of synthesis of the production amino acid may be determined by the method of Atfield and Morris described in the previous section.

A second method is to incubate the amino acid with a very small amount of 2-oxo acid and determine the product 2-oxo acid by formation of a dinitrophenylhydrazone. Alanine aminotransferase may be assayed by incubating the enzyme with 0.1 м alanine and 0.002 м 2-oxoglutarate in 0.1 м Tris/HCl buffer pH 7.4. The pyruvate formed may be determined by reaction with 2,4-dinitrophenylhydrazine and measuring the colour at 546 nm. A standard curve of varying pyruvate concentrations must, of course, be constructed.

A third more refined but expensive method is to couple the 2-oxo acid produced to NADH oxidation by an added enzyme. A standard reaction mixture may be set up containing 25 μl 10 mм 2-oxoglutarate, 20 μl 10 mм EDTA, 10 μl 10 mм NADH and 40 μl enzyme extract. If aspartate aminotransferase is to be measured the product is oxaloacetate which is rapidly converted to malate by malic dehydrogenase with the subsequent oxidation of NADH which may be measured on a spectrophotometer at 340 nm in a similar manner to glutamate dehydrogenase. The final reaction medium in the spectrophotometer cell should also include 25 μl 10 mм aspartate, 0.5 ml 0.1 м HEPES buffer pH 8.0, 0.1 ml of 100-fold diluted commercial malate dehydrogenase and 0.28 ml H$_2$O. In crude extracts of plants there is often sufficient malate dehydrogenase already present to drive the reaction without further addition of the enzyme. If alanine aminotransferase is to be measured pyruvate is the product, which may be converted to lactate by lactate dehydrogenase. In this case the final reaction medium should also include 0.1 ml 0.1 м alanine, 0.5 ml 0.1 м HEPES buffer pH 7.5, 0.1 ml of 100-fold diluted commercially available lactate dehydrogenase and 0.2 ml H$_2$O. The activity of the enzyme may be calculated in both cases using a blank reaction cuvette containing no 2-oxoglutarate.

8.4.8. Transport of Nitrogenous Compounds

If ammonia is formed in the roots it has to be transported to the leaves and developing fruits in a non-toxic form. Nitrate, on the other hand, may be transported directly in the xylem as the unmetabolized ion. Plants can apparently be divided into two groups on the basis of their nitrogen transport compounds. Cereals and temperate legumes (e.g. *Pisum* and *Vicia*) utilize the amides asparagine and glutamine and to a lesser extent arginine.

Asparagine   Glutamine   Arginine

Tropical legumes (e.g. *Glycine*, *Phaseolus* and *Vigna*) utilize the ureides allantoin and allantoic acid.

Allantoin

Allantoic acid

All the compounds above are characterized by a high N:C ratio. It is not clear yet why tropical legumes are able to synthesize large amounts of the ureides with a very high nitrogen content.

Nitrogen may be transported in either the phloem or the xylem. Techniques for determining the content are somewhat difficult; perhaps the easiest is the successive cutting of legume pod tips described by Pate *et al.* (1974) and Fellows *et al.* (1978).

The composition of the xylem sap can be readily determined, particularly in young plants grown in well-watered soil. The only essential requirements are a very sharp razor blade and a piece of clear plastic tubing 3–4 cm long, with an internal diameter the same as the thickness of the stem. The shoot is cut immediately above the surface of the soil and the tubing placed rapidly over the cut end. Sap can be readily seen to collect in the tubing after about 30 minutes and may be collected with a syringe or pipette. The amino acid content of the sap may be determined by spotting 10–20 μl onto a TLC plate or paper chromatogram as described in Section 4.2. Amino acids can be readily identified by spraying with ninhydrin, and quantitative determinations can be made by the method of Atfield and Morris (see also Section 4.3).

The ureide content of the sap can be deter-mined by measuring the amount of glyoxylate produced after hydrolysis. The sap should be heated in 0.05 M HCl for 2 minutes at 100°C and immediately cooled in ice. The hydrolysed samples should then be incubated with phenyl-hydrazine hydrochloride at a concentration of 0.66 mg/ml at 30°C for 15 minutes. The mixture should be cooled in a salt-ice bath and made up to 3 M HCl and 3.3 mg/ml potassium ferricy-anide. After allowing colour development the absorbance of the solutions can be read at 520 nm. Standard curves of allantoic acid must be prepared at least in duplicate before any determinations can be made. The above test is not given by allantoin itself only allantoic acid. It is, however, possible to hydrolyse allantoin to the free acid in alkali prior to the assay.

8.4.9. The Biosynthesis of Amino Acids

It is still generally accepted that only twenty amino acids are incorporated into protein. There may, however, be considerable post-translation modification (e.g. the formation of *N*-methyly-sine, 4-hydroxyproline and 4-carboxyglutamate). An outline of the pathways of amino acid biosynthesis is shown in Fig. 8.5.

It can be seen that the carbon skeletons required for amino acid synthesis are derived from the tricarboxylic cycle acids, in particular pyruvate, oxaloacetate and 2-oxoglutarate. The only amino acids that can derive carbon directly from photosynthetic CO_2 fixation are the aromatic amino acids from erythrose-4-phos-phate. However, chloroplasts are able to syn-thesize amino acids provided they are supplied with a suitable substrate.

A pathway that has been extensively studied at Rothamsted is the synthesis of lysine, threonine and methionine from aspartate (Fig. 8.6). The amino acids are frequently those that limit the quality of plant foodstuffs; in particular cereal seeds are deficient in lysine and legume seeds in methionine. The ability of chloroplasts to synthesize these amino acids can be readily demonstrated using intact chloroplasts prepared by the method described in Section 6.6.

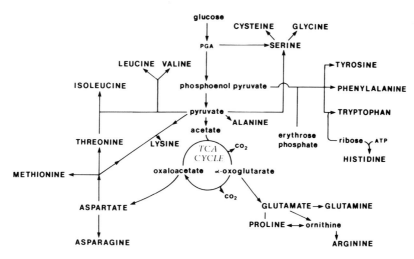

FIG. 8.5. The pathways of amino acid biosynthesis in plants.

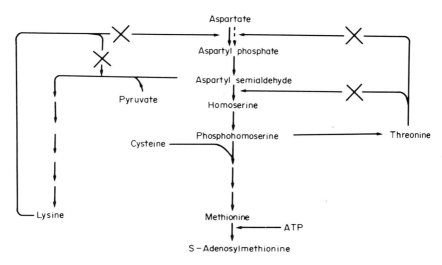

FIG. 8.6. The regulation of aspartate-derived amino acid synthesis.

Chloroplasts containing 100 μg chlorophyll should be resuspended in 0.5 ml buffer containing sorbitol (300 mM), EPPS (pH 8.3) and 30 mM KCl, with 2 μCi of ^{14}C-aspartate for 20 minutes in a bright light (controls should be run in complete darkness). The reaction is stopped by the addition of an equal volume of ice-cold 10% trichloroacetic acid. The reaction products can be separated by two-dimensional chromatography on TLC plates in a similar method to that described in Section 4.2. Amino acids can be identified by cochromatography with authentic standards.

Radioactivity in each amino acid can be determined by scraping each spot into a scintillation vial. A full description of this technique is given in Mills *et al.* (1980) and Davies and Miflin (1978). Pea leaf chloroplasts readily convert aspartate to lysine, threonine, homoserine and to a lesser extent methionine. This conversion is regulated by the endproduct amino acids. Lysine almost totally inhibits its own synthesis at levels above 1 mM, and also that of homoserine and threonine. Threonine, however, only inhibits its own synthesis and that of homoserine. The feedback inhibition can be readily explained by the known properties of the enzymes involved in the pathway. Aspartate kinase which catalyses the first step, the formation of aspartyl phosphate, is inhibited strongly by lysine and to a lesser extent by threonine. The inhibition by lysine and threonine is additive suggesting the presence of two separate enzymes. Lysine is also able to inhibit the first enzyme unique to its own synthesis dihydrodipicolinate synthase; and in a similar manner threonine inhibits homeserine dehydrogenase (see Fig. 8.6). The plant is thus able to carefully regulate the flow of metabolites along the complex pathways of amino acid synthesis. High levels of amino acids do not build up, and the rate of synthesis is then determined by the rate of incorporation into protein. A similar system operates in bacteria, where the regulation of metabolite flow is even more important. Bacteria are also able to regulate amino acid synthesis by "repressing" the amount of a particular enzyme synthesized within the cell. Although this may happen in the higher plants, there has, as yet, been no convincing evidence to demonstrate "repression and induction" for enzymes involved in amino acid metabolism.

The complex pathway to lysine, threonine, methionine and further on to isoleucine, leucine and valine requires a considerable amount of energy input in the form of NADPH and ATP. If these reactions are carried out in the chloroplast the energy is derived directly from light. The light dependence of the reactions can be demonstrated by keeping the reaction tubes in the dark, when little conversion of aspartate to other amino acids takes place. As these reactions utilize NADPH and ATP directly from the electron transport chain of the light process, they are true PHOTOSYNTHESIS in the same manner as CO_2 fixation. If amino acid synthesis is carried out in the root or maturing seed then the energy required for the synthesis must be derived from previously fixed carbon via oxidation in the mitochondria. It is possible to calculate the amount of glucose required to synthesize 1 g of methionine in the root from nitrate and sulphate as 2.13 g. If the reactions had gone on in the chloroplast using light energy then the cost would only be 0.51 g. Thus there is a considerable saving in energy if plants carry out their synthetic reactions in the chloroplast, and there could be a considerable saving in total yield if plants were selected which carried out more of their metabolism in the leaf rather than in the root or seed.

8.4.10. Selection of Amino Acid Metabolism Mutants

In Fig. 8.6 it can be seen that lysine and threonine are able to inhibit the passage of carbon from aspartate to homoserine by inhibiting both aspartate kinase and homoserine dehydrogenase. If they are added together to a growing plant they are able to kill the plant by preventing the formation of methionine due to the prevention of homoserine synthesis. Under normal conditions it is not possible to add lysine and threonine to growing plants: (1) due to the large costly amounts of the amino acids required to feed to plants growing in a pot or in the field; (2) due to the stimulation of bacterial and fungal growth which will remove the amino acids and interfere with the natural growth of the plants. This problem was initially overcome by growing plants in tissue culture, but it was not always possible to regenerate active plants after the tests had been carried out. Studies at Rothamsted over the last 5 years by Dr. S. W. J. Bright have developed a technique for growing young barley plants in sterile culture, in such a way that the action of various compounds can be tested, and viable plants can be obtained at the end of the experiment.

Method

Barley seeds should be dehusked in 50% v/v H_2SO_4 for 3 hours, washed in tap water and three changes of distilled water before being soaked overnight at 5°C. A wash in saturated $CaCO_3$ solution for 20 minutes after the acid treatment is also recommended. The embryos may then be dissected by hand from the seeds and allowed to dry on filter paper at 25°C for at least 24 hours. Petri dishes containing agar (6 g/l), Murashige and Skoog's medium (*Physiol. Plant.* **15**, 473–497, 1962) without hormones or casein hydrolysate and sucrose 30 g/l which has been previously autoclaved should be prepared. If amino acids are to be tested they should be sterile filtered and added to the mixture whilst it is still liquid; 20 ml of medium may be added to a standard 9-cm plastic petri dish.

Dry embryos should be sterilized by shaking for 15 minutes at 25°C in the supernatant of 70 g/l calcium hypochlorite with a small amount of Teepol added as a wetting agent. The embryos are then rinsed in sterile distilled water, washed for 5 minutes in sterile 0.01 M HCl and finally rinsed with 400 ml sterile distilled water. Embryos should be placed on the agar medium under sterile conditions so that the scutellum faces downwards. The plates should be incubated for 7 days at 25°C under 8000 lux white lights with a 16-hour day. The embryos develop roots and leaves, and examples of the plants at different stages of growth can be seen in Fig. 8.7. After the leaves reach 6 cm long the plants may be successfully transplanted to a small pot of compost and, provided they are maintained in a humid atmosphere for a further period until the roots establish, they will grow into healthy mature plants.

If the plants are grown in the presence of 2 mM lysine and 2 mM threonine then there is an 80% reduction in growth. The inhibition in growth can be prevented by the addition of 0.5 mM methionine to the medium, suggesting that the action of lysine + threonine is to prevent methionine synthesis (see Fig. 8.8).

If normal embryos are grown in the presence of lysine and threonine then none of them will show any significant leaf development. However, if the seeds are first treated with a mutagen then there is a large increase in the probability of obtaining a mutant plant that is in some way resistant to the toxic action of lysine + threonine. In Fig. 8.8 the presence of such a plant can be seen, after azide treatment. A number of these mutants in barley have now been obtained at Rothamsted and their enzymology examined. It would appear that in at least one case the plant synthesizes a modified form of aspartate kinase that is less sensitive to feedback inhibition by lysine, and thus allows carbon to flow along the pathway for methionine synthesis. (The work on this project was carried out by Dr. S. E. Rognes and Dr. Bright.) The ultimate aim of the mutant selection work is to produce barley plants that have enzymes in the pathway, with altered regulatory characteristics and are able to accumulate high levels of essential amino acids. These plants would thus be of considerable value in improving the nutritive quality of cereal protein.

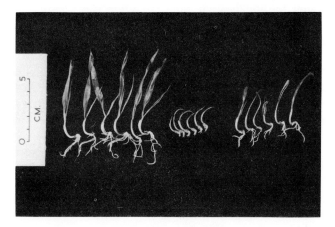

FIG. 8.7. Different stages of barley embryo development grown in sterile culture. The plants on the left are 7 days old and are the maximum size that can be grown.

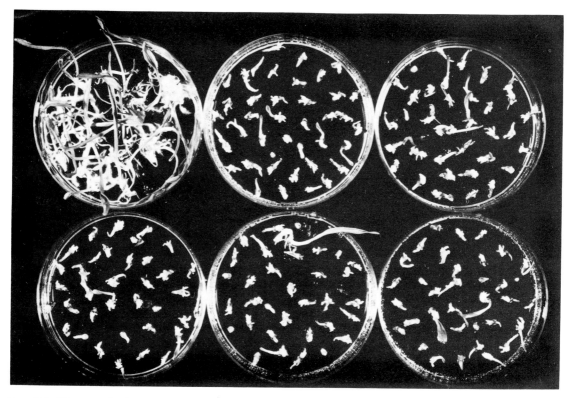

FIG. 8.8. The growth of barley embryos in the presence of 2 mM lysine and 2 mM threonine. The top left plate also contains 0.5 mM methionine. Note in the bottom centre plate, the presence of one plant that is growing normally.

Mutagen Treatment Technique

The treatment described, although used on barley, can probably be used for most dry seeds. A simple test as to whether a mutagen has acted is the appearance of a high proportion of "chlorophyll less" or albino plants, when the seeds are germinated (5–10%).

Seeds should be soaked overnight in water at 4°C, and then incubated in 1 mM sodium azide in 0.1 M phosphate buffer pH 3.0 for 2 hours. The seeds should then be washed in two volumes of distilled water and under running tapwater in the cold for 30 minutes. The seeds should be dried and then may be planted in the field or in pots. The seeds that have been harvested after the growth of these plants (M2 generation) should be used for the embryo-growth experiments discussed above.

Warning

Azide is a volatile mutagen and a respiratory poison, coupled with which it may form an explosive mixture if poured down the sink. *ALL* handling of azide must be carried out in a fume cupboard, the original solution and the first two washings must be treated with a 15% solution of ammonium ceric nitrate before they are poured down the sink, after this treatment the azide is decomposed.

The method is perfectly safe as long as reasonable precautions are taken.

Bibliography and Further Reading

ATFIELD, G. N. and MORRIS, C. J. O. R. (1961) Analytical separation by high voltage electrophoresis-amino acids in protein hydrolysates. *Biochem. J.* **81**, 606.

BEEVERS, L. (1976) *Nitrogen Metabolism in Plants.* (Edward Arnold.)

DAVIES, H. M. and MIFLIN, B. J. (1978) Advantage of *o*-phthaldehyde for visualizing ^{14}C labelled amino acids on thin layer chromatograms and an improved method for their recovery. *J. Chromatog.* **153**, 284.

FELLOWS, R. J., EGLI, D. B. and LEGGETT, J. E. (1978) A pod leakage technique for phloem translocation studies in soybean (*Glycine max* L. mer.). *Plant Physiol.* **62**, 812.

HEWITT, E. J. and CUTTING, C. V., eds. (1979) *Proceedings 6th Long Ashton Symposium on Nitrogen Assimilation.* (Academic Press: London.)

MIFLIN, B. J., ed. (1980) *Biochemistry of Plants*, Vol. 5. (Academic Press: London, N.Y.)

MIFLIN, B. J. and LEA, P. J. (1976) The pathway of nitrogen assimilation in plants. *Phytochem.* **15**, 873.

MIFLIN, B. J. and LEA, P. J. (1977) Amino acid metabolism. *Ann. Rev. Plant Physiol.* **28**, 299.

MILLS, W. R., LEA, P. J. MIFLIN, B. J. (1980) Photosynthetic formation of the aspartate family of amino acids in isolated chloroplasts. *Plant Physiol.* **65**, 1166.

PATE, J. S., SHARKEY, P. J. and LEWIS, O. A. M. (1974) Phloem bleeding from legume fruits—a technique for study of fruit nutrition. *Planta* **120**, 229.

RONZIO, R. A., ROWE, W. B. and MEISTER, A. (1969) Studies on the mechanism of inhibition of glutamine synthetase by methionine sulfoximine. *Biochemistry* **8**, 1066.

WIGHTMAN, F. and FOREST, J. C. (1978) Properties of plant amino transferases. *Phytochem.* **17**, 1455.

ISOLATION OF ENZYMES

J. COOMBS

9.1. INTRODUCTION

Enzymes are naturally occuring macromolecules (globular proteins) which are constituents of the cells of all living matter. They are polymers of α-amino acids (I) joined by peptide linkages formed by elimination of a molecule of water from the carboxyl group of one amino acid and the amino group of the next to form a peptide linkage (II).

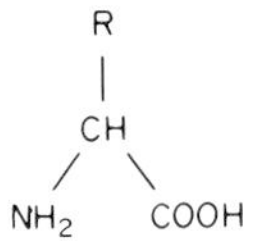

R
|
CH
／　＼
NH$_2$　　COOH

(I)

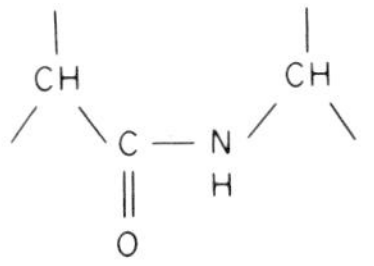

CH　　　　CH
／　＼　／　＼
　　C — N
　　‖　　H
　　O

(II)

All proteins are made up of about twenty naturally occurring amino acids, differing in the group represented as R in (I) above. This may be acidic, basic, or neutral, aliphatic or aromatic. Each protein has its own characteristic sequence of amino acids. In all natural amino acids, other than glycine in which R=H, the central carbon atom is asymmetric which means that the amino acid may occur in either the D or the L form. Most natural amino acids are in the L form.

A given enzyme may consist of one or more polypeptide chains wound into a specific three-dimensional structure. This structure, which is essential for the catalytic activity, is maintained by hydrogen bonding, electrostatic forces, disulphide bridges and hydrophobic interactions so that most of the polar groups are at the surface and most non-polar groups are sited towards the interior of the molecule.

The main function of proteins that we consider here is their role as catalysts. If we consider the reversible reaction:

$$A + B \rightarrow C + D$$

then the equilibrium constant (K) is defined as shown below:

$$K = \frac{[C][D]}{[A][B]}$$

where [A], [B], [C] and [D] are the concentrations of the various compounds at equilibrium. An enzyme, or indeed any other catalyst, can only promote such a reaction if there is an overall loss of free energy—it cannot alter the thermodynamic characteristics of the reaction. In many cases the reaction may be essentially irreversible. However, such a conversion may be driven in the opposite direction by coupling it with a second reaction in which there is a greater loss of free energy, so that on balance there is a small net loss of free energy. In either case the role of the enzyme is to speed up the reaction and decrease the time in which the reaction reaches equilibrium. The rate at which the enzyme catalyses the conversion of one or more compounds (substrates) to the desired end product depends on a number of factors as detailed below.

9.2. ENZYME ACTIVITY

The major factors which affect the rate at which an enzyme catalysed reaction will proceed are shown in Fig. 9.1. These factors include the concentration of enzyme, substrate (and co-factors, metal ions, etc.) and possible inhibitors as well as physical or environmental conditions such as pH, temperature and ionic strength. These factors are now considered briefly.

(a) *Protein concentration*. In general the rate of reaction observed *in vitro* in an enzyme assay will be proportional to the protein concentration, as long as all other parameters are optimized. However, at high protein concentrations the rate may be limited as the protein may form aggregates with lower catalytic activity on a protein basis. Protein may also complex with other compounds in the reaction mixtures, or part of the protein may be inactivated by trace impurities in other reagents used. This can result in part of the enzyme not acting as a catalyst, which may be indicated if

activity is plotted against protein concentration, by an intercept on the protein concentration axis proportional to the amount of inhibitor or impurity. Hence, it is essential to check the proportionality between rate and protein concentration for any new assay or protein preparation.

(b) *pH*. Most enzymes work within the pH range of 4.5 to 9.5. Some show very marked pH optima, others may be less specific working over a wide range. When looking at a new enzyme, or an enzyme from a new source it is important to establish the pH optima. Conventionally, enzyme activity is determined at this optimum pH. However, in many cases, especially with regulatory enzymes, they may not in fact operate at this pH *in vivo*, and a greater degree of metabolite control may be shown at pH values below that required for maximum catalytic activity.

(c) *Time*. If the concentration of substrate is kept constant and the products of the enzyme reaction removed then the rate of reaction with a stable enzyme should be constant with time. However, most enzyme assays are of the batch type, i.e. a reaction mixture containing a fixed amount of substrate is measured for a fixed time. In this case the actual rate of reaction may drop as the substrate is consumed, and in some cases the products of the reaction may in fact act as inhibitors. It is therefore important to establish that the reaction is linear over the time period used for assay, or alternatively to determine the detailed time course in order to establish the "zero time" velocity (obtained by drawing a tangent to the curve at zero time). In general the reaction will depart significantly from linearity at potentially limiting substrate concentrations when 15% of the substrate has been used. This possibility must be checked by calculation from the rate of substrate use, the time of assay and the initial concentration of substrate. Particular care must be taken in assays where low concentrations of substrates have been used, for instance in serial dilutions used to establish kinetic data. If a fixed time assay is used the calculated rate at low substrate concentrations may be erroneously low, giving the appearance of a sigmoid substrate/rate relationship.

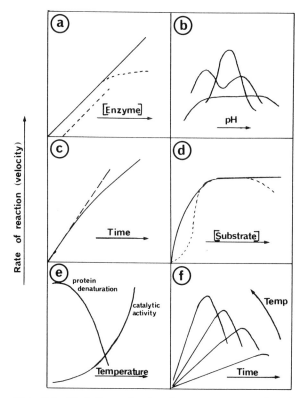

FIG. 9.1. Variation of rate of an enzyme catalysed reaction with changes in various parameters as indicated.

(d) *Substrate concentration.* The response of enzymes to variations in substrate concentrations is considered in more detail in the appendix on enzyme kinetics. However, in general the change in conversion of substrate to product with increasing substrate concentration will show two phases. In the first phase (low substrate concentration) the rate of conversion of substrate to product is proportional to the enzyme activity, whereas during the second phase (saturated) the rate of reaction is independent of substrate concentration. When an experiment is designed to establish the level of a given enzyme in an extract of some tissue the reaction medium is formulated in such a way that it is not limiting in respect of either substrate, co-factors or essential metal ions. Once again if a new plant material is being investigated care should be taken to establish that the activity is not in fact inhibited by high substrate concentrations. Some enzymes do not show a linear increase in activity at low concentration of substrates, but exhibit a biphasic or multiphase (sigmoid) response. In many cases such enzymes play an important regulatory role *in vivo.* However, there are many ways of obtaining artefacts which give sigmoid kinetics. Hence, considerable care must be taken before a role in metabolic control is assigned to such an enzyme. Such considerations lie outside the scope of the present discussion. Students encountering such anomalous kinetics are urged to discuss their work with an experienced worker in this field before jumping to too many conclusions.

(e) *Temperature.* It is particularly important that all assays are carried out at a carefully controlled temperature. The response of an enzyme to temperature can be considered in terms of two components. First, there is the exponential increase in catalytic activity (approximately doubling for every 10° rise in temperature) and secondly there is a decrease in catalytic activity resulting from part of the enzyme being denatured by heat. This heat denaturation increases as the temperature is increased. The result of the interaction of these two components are shown in Fig. 9.1(f). Care should be taken in the design of experiments set up to compare the temperature profile of enzymes isolated from different varieties or strains of plants which exhibit different degrees of resistance to hot or dry climates.

9.3. ISOLATION OF ENZYMES

The isolation and preparation of an enzyme involves the following considerations or steps:

(a) Aim—define purpose of isolation.
(b) Select material—pretreatment, e.g. light/dark chilling may be necessary. Wash—has it been sprayed with chemical inhibitors.
(c) Disrupt cells in suitable media.
(d) Precipitate protein.
(e) Purify enzyme—increase specific activity.
(f) Assay.

As far as higher plants are concerned there are a number of particular problems to be faced in the isolation and purification of an enzyme. These include (1) the low protein concentration of many plant tissues; (2) the presence of inhibitory compounds such as acids, phenols or ions (often in the vacuoles) which are liberated at the same time as the protein when the cells are broken; (3) the tough cell wall (ligno-cellulose) which requires a powerful (possibly heat generating) technique to disrupt it; (4) the cells contain many different enzymes each at a low concentration, or constituting only a small fraction of the total cell protein. An exception to this may be the concentration of ribulose-bisphosphate carboxylase in photosynthetic tissue.

9.4. PROTEIN DETERMINATION

During the isolation of an enzyme it is necessary to be able to monitor the purity of the enzyme by determining the protein level in any extract or sample. The various common methods of protein determination are summarized below. As a result of knowing the protein concentration and catalytic activity of a given sample or preparation the specific activity may be calculated. This may be defined as the units of activity per mg of protein. One unit is that amount of enzyme which will catalyse the transformation of one micromole of substrate per minute under standard conditions. However, any value must be regarded with caution since protein purity may not be the same as active site purity.

Kjeldahl Method

This is a measurement of the protein nitrogen content. A sample containing approximately 1 mg of nitrogen (approximately 6 mg of protein) is digested with concentrated sulphuric acid (2 ml) in a long-necked flask with 1 g of a catalyst mixture (prepared by grinding to a powder in a mortar 80 g K_2SO_4, 20 g $CuSO_4 . 5H_2O$ and 0.34 g of sodium selenate). After gentle refluxing for about 8 hours, during which time the protein nitrogen is quantitatively converted into ammonium sulphate, the flask is allowed to cool, and the contents carefully diluted with about 25 ml of distilled water.

The solution is then transferred quantitatively to a distillation apparatus with an outlet tube immersed in a flask containing boric acid (A.R.) solution (2%, 10 ml). Sodium hydroxide solution (30%, 10 ml) is then added to the diluted digestion products in the distillation chamber. Steam is passed through the mixture and the ammonia liberated from the ammonium sulphate under the alkaline conditions distils over into the boric acid solution. About 20 ml of distillate is collected. The absorbed ammonia is then determined by titration with standard hydrochloric acid solution (0.01 M) using as indicator a solution of methyl red (0.2%) and methylene blue (0.1%) in ethanol, which shows a colour change from purple to grey when all the alkali has been titrated. It is usual to run a control of ammonium sulphate. 1 ml of 0.01 M HCl corresponds to 0.14 mg of nitrogen.

This method measures total nitrogen, whether in protein or otherwise. For this reason contaminating nitrogenous substances such as ammonium sulphate must be absent from the sample. Most proteins have a nitrogen content of about 16% and thus to convert the nitro-

gen value into a weight of protein, multiply by 6.25.

Biuret Reaction

In strongly alkaline solution all compounds containing two or more peptide bonds react with copper salts to give a violet colour. The reaction is used as the basis of a qualitative test for protein and when quantified provides a useful colorimetric measurement of protein.

The biuret reagent is prepared by dissolving 1.5 g of $CuSO_4·5H_2O$ and 6.0 g of sodium potassium tartrate ($NaKC_4H_6O_6·4H_2O$) in distilled water (500 ml). Sodium hydroxide solution (10%, 300 ml, carbonate free) is then added to this solution with constant stirring, followed by dilution of 2 litres with distilled water. The reagent should keep indefinitely if stored in a polyethylene bottle.

To 1 ml of protein solution, containing from 1–10 mg of protein, is added 4.0 ml of biuret reagent and the whole immediately mixed vigorously. After standing 30 minutes at room temperature the optical density of the solution is read at 550 nm. A calibration graph is prepared using known amounts of a suitable protein such as bovine serum albumn.

Lowry Method

This is the most widely used method for protein determination. The Folin-Ciocalteau phenol reagent reacts with certain portions of the protein molecule producing a dark-blue colour.

Reagents are prepared as follows:

A. Anhydrous sodium carbonate (2%) in sodium hydroxide (0.1 M).
B. Copper sulphate solution (0.5%).
C. Sodium potassium tartrate solution (1%).
D. 48 ml A + 1 ml B + 1 ml C mixed immediately prior to use.
E. Folin-Ciocalteau reagent diluted with water (usually 1:1) to give a solution 1 N in acid.

A sample (0.5 ml) containing up to 500 μg of protein is mixed with 5.0 ml of reagent D. The solution is allowed to stand 15 minutes at room temperature. Reagent E (0.5 ml) is then added and mixed at once. After standing for 30 minutes at room temperature the optical density is read at 700 or 750 nm. It is necessary to prepare a calibration graph using known amounts of bovine serum albumin as standard since the colour produced is not directly proportional to the amount of protein present.

9.5. EXTRACTION OF PROTEINS

The method used for the extraction of protein from a given tissue depends to a large extent on the nature of the starting material. For efficient extraction it is necessary to disrupt the cellular structure. This may be done using a pestle or mortar, for small samples, or using a homogenizer or blender with rotating knife blades for larger samples. Due to the possible generation of heat the blender vessel should be cooled on ice or in a refrigerator prior to use, and a cooled medium should be used. The resulting brei is then filtered through several layers of muslin or cheese-cloth, and smaller debris removed by centrifugation. In some cases the required protein may in fact be complexed with other cell constituents and thus a proportion lost during the centrifugation step. Sometimes it is possible to solubilize such protein by sonication, freeze-thawing or use of detergents such as Triton-X100.

The method used to obtain the best protein preparation from a given tissue is often determined in an empirical manner, the best isolation medium being designed by changing pH, ionic strength and the nature of additives until the best result is obtained. The best conditions may be determined by measuring both the amount of enzyme activity detectable, and the level of protein released in any given procedure.

9.6 PROTECTION FROM PHENOLS AND PHENOL OXIDASE ACTIVITY

Many plants have very high levels of inhibitory substances which are released at the same time as the protein. In particular o-diphenols are a problem. In the presence of o-diphenol oxidase usually present in the same cells the diphenol is converted to the quinone. This reacts with —SH or —NH$_2$ groups with the following effects:

1. Inactivation of the enzyme.
2. Precipitation of protein – i.e. make soluble enzymes appear particulate.
3. Change physical characteristics such as redox potential, UV absorbtion spectra.

To reduce effects to a minumum

1. Keep preparations cold, 0 to 4°C.
2. Remove O$_2$ (i.e. under N$_2$).
3. Add inhibitors (copper chelators, e.g. diethyl dithiocarbamic acid).
4. Add reducing compounds, e.g. ascorbate.
5. Add thiols or mercapto compounds, e.g. thioglycollate, mercapto-ethanol, dithiothreitol.
6. Add an excess of another protein, e.g. BSA.
7. Add polymers which complex phenols, e.g. PVP polyethylene glycol, polyamid, nylon 66.

The level of phenol oxidase in a plant extract, and the efficiency with which it is combated using these compounds can be followed using the oxygen electrode (Baldry et al., 1970).

Experiment. Assay and inactivation of phenol oxidase.

Materials. 10 mM chlorogenic acid, catechol or DOPA, oxygen electrode, pestle/mortar or blender, DIECA, thiols, mercaptans, polymers, made up in 10 g concentration series from 10 mM to μM.

Method. Add 2.3 ml water to electrode chamber, add heat extract, add inhibitors to give 3 ml total reaction mixture. (Repeat grinding with BSA, PVP, etc.) Initiate reaction by addition of 0.1 ml phenol as substrate.

Results. Plot observed rates as a function of concentration of added inhibitor.

From such experiments, conditions where oxidase is reduced to a minimum can be obtained and used to help in the design of a grinding medium.

9.7. CRUDE EXTRACTS

A large number of physiological studies are reported in which the activity of a crude enzyme preparation from one species, variety or tissue of a plant is compared with that from another. Although such an approach may yield useful data it is important that certain precautions are taken. Problems may arise due to differences in levels of endogenous inhibitory compounds such as phenols, organic acids or salts present in the various samples, or to differences in the levels of co-factors or substrates in the extracts. In crude extracts a wide variety of enzymes may be present hence the reaction (or change in measured parameter) which is seen may not be the one you wish to observe. These problems can be overcome to some extent if care is taken in the design of the isolation medium and some partial purification is carried out by either gel filtration on Sephadex® G25 (see Section 9.9) or by ammonium sulphate precipitation and dialysis (9.8).

9.8. AMMONIUM SULPHATE PRECIPITATION

This technique (salting out) is based on the fact that the solubility of most proteins decreases at high electrolyte concentrations. It is the anion in the added salt which is most important, and multivalent ions are more effective than monovalent ions such as chloride. Sulphate is most widely used, in the form of ammonium sulphate. Fractionation, using this salt, is one of the most useful procedures available in the early stages of protein purification, especially where large volumes of solution are being used (often a necessity when using plant tissue—resulting from low protein content). The salt may be added as a saturated solution or as the solid, with stirring in a cold room or in a beaker on ice. It is advisable to check the pH at intervals, adding ammonia in order to keep the pH close to neutrality. Various proteins will precipitate at different specific salt concentrations—the concentration being expressed as a percent saturation. For an unknown enzyme it is usual to add salt to 25% saturation, collect the precipitate by centrifugation, then collect further fractions after increasing the salt concentration by 10% saturation intervals to a final 80% saturation. A simple table is shown (Table 9.1) which indicates the amount of ammonium sulphate to add to obtain the required saturation. A more detailed and accurate table appears in Wood (1976).

The precipitated protein samples are dissolved and the salt removed by dialysis against at least five changes of distilled water, or suitable buffer (it is often advisable to add Mn or Mg chloride to the dialysis buffer). A small column of Sephadex G25 may also be used to desalt the protein. The samples may then be analysed for protein and the required enzyme activity, the specific activity of the various fractions determined and those of higher activity combined and used directly or purified further. Thereafter, a single protein cut can be taken by adding salt to the lower % saturation, discarding the precipitate, then adding salt to the higher % saturation limit and collecting the resulting precipitate. During the dialysis the volume may become quite large. If apparatus is not available for freeze-drying or ultrafiltration the protein can be concentrated by placing the dialysis tube in a beaker and packing the beaker with solid sugar. If left overnight, in the cold, much of the water will pass from the dialysis tubing due to osmosis.

Hints on dialysis. Knot one end of the tube as supplied by wetting, twisting into a "rope" and tie at least three knots. Add the enzyme solution from a pipette. If the tube is difficult to open, slide the point of a hypodermic needle into the unknotted end, open slightly, then blow down the tube to inflate the full length. Add enzyme solution to fill not more than about one-third. Expel air, twist top and knot as before, leaving about two-thirds of the tube empty and flat. This will allow the solution to expand as it takes up water by osmosis. *Note*: any marking made on the tube with a pen will probably come off. Labels may be tied on. However, a useful technique is to add a specific

Table 9.1. *Data for ammonium sulphate fractionation*

Initial concentration of ammonium sulphate, % saturation	Final concentration of ammonium sulphate, % saturation																
	10	20	25	30	33	35	40	45	50	55	60	65	70	75	80	90	100
	Grammes solid ammonium sulphate to be added to 1 litre of solution																
0	56	114	144	176	196	209	243	277	313	351	390	430	472	516	561	662	767
10		57	86	118	137	150	183	216	251	288	326	365	406	449	494	592	694
20			29	59	78	91	123	155	189	225	262	300	340	382	424	520	619
25				30	49	61	93	125	158	193	230	267	307	348	390	485	583
30					19	30	62	94	127	162	198	235	273	314	356	449	546
33						12	43	74	107	142	177	214	252	292	333	426	522
35							31	63	94	129	164	200	238	278	319	411	506
40								31	63	97	132	168	205	245	285	375	469
45									32	65	99	134	171	210	250	339	431
50										33	66	101	137	176	214	302	392
55											33	67	103	141	179	264	353
60												34	69	105	143	227	314
65													34	70	107	190	275
70														35	72	153	237
75															36	115	198
80																77	157
90																	79

coloured bead to each tube before closing the top.

The resulting dialysed protein solution may be used for crude assays. However, for some applications a purer preparation may be required. The most common, relatively simple techniques now available for such purification are ion exchange and gel filtration chromatography.

9.9. MODERN METHODS OF PROTEIN PURIFICATION

(a) Ion Exchange Chromatography

The most frequently used cellulosic ion-exchangers are diethylaminoethyl (DEAE-) cellulose and carboxymethyl (CM-) cellulose, which are weak-base (anion) and weak-acid (cation) exchangers respectively. These are prepared from cellulose by substitution of some primary and secondary hydroxyl groups of the anhydroglucose units with diethylaminoethyl- and carboxymethyl- groups joined through ether linkages.

These materials, prepared in such a way as to leave most of the crystalline structure in the cellulose intact, are relatively hydrophilic. They have an open structure readily penetrated by large molecules and a large surface area. Furthermore, they have a high capacity for absorption of protein, the binding is often freely reversible, mild conditions being required to effect desorption. Columns of these materials are readily prepared and can be run at high flow rates (e.g. 100 ml/hr).

DEAE-cellulose is preferred for chromatography of acidic and slightly basic proteins at pH values above their isoelectric points since at such pH values the DEAE group is highly ionized. The most useful range is pH 6–8. At higher pH values it is in the uncharged form and will not bind proteins. At lower pH values most proteins (apart from the very acidic ones) have their negative charges neutralized and will be bound.

CM-cellulose is useful for chromatography of neutral and basic proteins, i.e. those which are positively charged above pH 4. Below this pH few of the carboxymethyl groups are ionized.

The principle of ion-exchange chromatography depends on the binding of protein to an ion-exchanger carrying a charge of opposite sign. Careful control of pH is important. Most common proteins, which have isoelectric points in the region of 5–6, are negatively charged at pH 7.5 and will bind to DEAE–cellulose at this pH. They will be positively charged at pH 4.5 and bind to CM–cellulose. It is generally pointless to carry out chromatagraphy on CM–cellulose at high pH or on DEAE–cellulose at low pH unless it is suspected that the desired protein is either very basic or very acidic.

Desorption of proteins from ion-exchange celluloses is accomplished by changes in pH or ionic strength. Desorption by change in ionic strength is usually brought about by incorporation of salt (e.g. NaCl) into the eluting buffer. As the electrolyte concentration is increased, protein is displaced from the ion-exchanger by the salt anions (DEAE–cellulose) or cations (CM-cellulose). The most weakly held species are eluted at low salt concentration and more tightly bound proteins at higher concentrations.

Most cellulosic ion-exchangers are supplied in the form of a dry fibrous powder. Before use it should be washed repeatedly with 0.5 M sodium hydroxide solution, with filtration on a Buchner funnel between washings. When the washings are clear, excess alkali is removed by washing with water. This is followed by a treatment with 0.5 M hydrochloric acid solution, filtration, and a further water wash to remove excess acid. This precedure converts CM–cellulose into the uncharged (—COOH) form and DEAE-cellulose into the charged form. The latter is then converted into the free base by suspending it in 0.5 M sodium hydroxide solution, followed by filtration and a water wash. In this way the ion-exchangers are obtained in the forms which, theoretically at least, are most easily equilibrated with starting buffer, as follows.

The ion-exchanger is suspended in the required buffer and adjusted to the pH at which the material is to be used (by slow addition of the acid (for DEAE-cellulose) or base (in the case of CM-cellulose) component of the buffer). The exchanger equilibrated in this way is then filtered and washed with several volumes of the chosen starting buffer until the washings have the same pH and conductivity as this buffer. The required amount of material, in the form of a slurry, is poured *into* a column partially filled with buffer. Buffer is then allowed to flow from the column, addition of the exchanger being made till the column is filled to the required height. The uniformity of packing of such a column is not nearly as critical as for molecular-sieve chromatography columns, and as a result they may be packed quickly and simply without any special precautions other than to ensure that the bed does not run dry at any time. The column should be eluted with a few bed-volumes of starting buffer before application of sample.

(b) Molecular-sieve Chromatography

Columns packed with certain swollen gels separate molecules on the basis of molecular size. This method of fractionation which has been used extensively in protein purification depends on the fact that molecules of different sizes vary in their ability to penetrate the swollen gel particles. Some molecules, of high molecular weight, are excluded from the gel matrix. Smaller molecules penetrate the gel to a greater or less extent. Thus high-molecular-weight proteins excluded from the gel are eluted from the column first. Low-molecular-weight compounds are eluted last because, being able to penetrate the gel, the fraction of the total column volume available to them is greatest. Molecules with intermediate molecular weights, which have limited abilities to penetrate the gel particles, are eluted at volumes between these two extremes.

Table 9.2. *Materials for molecular-sieve chromatography*

Molecular sieve	Approximate exclusion limit	Useful fractionation range
Cross-linked dextrans		
Sephadex G–25	5000	1000–5000
Sephadex G–50	30,000	1500–30,000
Sephadex G–75	70,000	3000–70,000
Sephadex G–100	150,000	4000–150,000
Sephadex G–150	400,000	5000–400,000
Sephadex G–200	800,000	5000–800,000
Cross-linked polyacrylamides		
Biogel P–6	6000	1000–6000
Biogel P–10	20,000	1500–20,000
Biogel P–30	40,000	2500–40,000
Biogel P–60	60,000	3000–60,000
Biogel P–100	100,000	5000–100,000
Biogel P–150	150,000	15,000–150,000
Biogel P–200	200,000	30,000–200,000
Biogel P–300	400,000	60,000–400,000
Agarose gels		
Sepharose 6B	4×10^6	..
Sepharose 4B	20×10^6	..
Sepharose 2B	40×10^6	..

The most commonly used materials for this purpose are cross-linked dextrans (Sephadex), cross-linked agarose (Sepharose) and cross-linked polyacrylamide (Biogel). Sephadex and Biogel are supplied in the form of dry beads which must be swollen before use; Sepharose is supplied ready-swollen. These materials are available with varying degrees of cross-linking and in a series of particle sizes. The extent of cross-linking determines the size of the pores in the gel and hence the fractionation ranges of the various molecular sieves (Table 9.2).

When the material is supplied as dry beads, these must be fully hydrated before use. This is done by allowing them to swell in water solution at room temperature with gentle stirring. Alternatively, they may be prepared for use rather more quickly by heating the suspension in a boiling-water bath for a few hours. The swollen particles are then allowed to partially settle in a beaker and decanted several times to remove fines.

Precautions must be taken to avoid microbial contamination on columns.

The sample, equilibrated with the required buffer, should be layered onto the top of the column and allowed to sink into the column under the influence of gravity. When application of the sample is complete, elution should be continued with 1–2 bed-volumes of starting buffer to remove any unadsorbed material. Assuming that the conditions have been such as to effect adsorption of the required protein, the elution conditions should then be changed so as to fractionally desorb the proteins from the column. This is done, as has been already mentioned, by increasing the ionic strength of the eluent or alternatively by changing its pH but keeping the ionic strength constant.

Gradient elution is the superior method, giving better resolution of the desorbed proteins. Linear gradients are the most commonly used. However, concave salt gradients are particularly useful since these give a relatively shallow gradient in the range in which most proteins are eluted from the ion-exchanger, but still reach a final concentration high enough to bring about elution of tightly bound proteins.

Use of these ion-exchangers in a batchwise fashion is also of value. This method is especially useful for large-scale operations. For example protein may be adsorbed from a large volume of dilute solution by stirring with ion-exchange cellulose followed by filtration. Care has to be taken to ensure that the solution is at a suitable pH and ionic strength for adsorption to take place efficiently. The protein may be recovered from the filter cake by washing with a small volume of an appropriate solution (e.g. 1 M sodium chloride). This procedure brings about concentration to a more workable volume.

Elution with a few bed-volumes of 0.02% sodium azide will help to prevent the development of microbes.

After swelling and removal of fines, the gel suspension is degassed by placing it under vacuum until evolution of dissolved air ceases (10–15 minutes). It may then be used to pack a suitable column, the most useful sizes for preparative work being 2.5×100 cm or 5.0×100 cm. The recommended method of packing involves partly filling the column with the chosen eluent solution, into which the gel slurry is poured. When a layer of gel about an inch deep has settled at the bottom of the column, solution is allowed to flow out and more gel added until the column is full. During this process the gel in the column must not be allowed to settle completely before the addition of more slurry as this will result in "layering".

At a constant flow rate the volume needed to elute the required enzyme may be determined by collecting samples from the column and assaying for the required activity. This volume (V_e) is not a reliable parameter as it varies with the total volume of the packed bed (V_t) and with the way in which the column has been packed. The elution of the enzyme is thus best characterized by a distribution coefficient (K_d) where $K_d = V_e - V_0/V_s$ and $V_0 =$ void volume, $V_s =$ volume of the gel. In practice V_s is difficult to measure. Hence, it is easier to substitute V_s by $V_t - V_0$ to give $K_{av} = V_e - V_0/V_t - V_0$, such that K_{av} represents the fraction of the stationary gel volume which is available for diffusion of a given solute species. If the ratio V_e/V_0 is plotted against log of molecular weight a linear relationship will be found.

9.10. SEPARATION OF PEP CARBOXYLASE

1. Blend 100 g de-ribbed leaf lamina in suitable media (200 ml), remove large cell debris by filtration through muslin.
2. Centrifuge to remove smaller debris.
3. Take supernatant.
4. Add ammonium sulphate to 30% saturation, check pH.
5. Leave 3 hours—discard ppt.
6. Add ammonium sulphate to 60% saturation.
7. Adjust to pH 7. Leave 3 hours.
8. Take up in distilled water.
9. Dialyse against at least four changes (litres) of distilled water.
10. Add to 50 ml DEAE cellulose in 0.05 M Tris HCl pH 7.0.
11. Pellet—discard supernatant.
12. Wash 0.1 M Tris HCl pH 7.0.
13. Discard supernatant.
14. Elute 0.2 M Tris HCl.
15. Add 60% ammonium sulphate at pH 8.0.
16. Leave, harvest ppt, dialyse, or desalt on small Sephadex G-25 column.
17. Concentrate by reverse dialysis against sucrose (crude PEP carboxylase).
18. Apply to calibrated column of G–200.
19. Collect fractions concentrate.

Throughout assay enzyme activity and protein (using Lowry assay).

Use crude PEP carboxylase to study enzyme kinetics (see Appendix)

9.11. CARBOXYLASE ASSAYS

1. ^{14}C. Reaction volumes of 0.5 ml containing 5 mM concentration of PEP and $MgCl_2$, 0.05 M HEPES buffer pH 8.0, and about 0–1 units of PEP carboxylase pre-incubated in a water bath. Reaction initiated by addition of $H^{14}CO_3^-$ (specific activity 1 $\mu Ci/\mu mol$) and 20-μl samples are taken at 10-second intervals. Each sample is injected into 200 μl of saturated dinitro-phenylhydrazine in 2 M HCl, mixed, left in cold, dried on planchettes and counted or counted on a liquid scintillation counter.

2. NADH linked. Reaction mixtures containing $MgCl_2$, 10 mM; PEP, 2 mM; NADH, 0.14 mM; malate dehydrogenase 5 units in 1 ml of Tris HCl pH 8.0. Initiate by addition of HCO_3^-, 5 mM. Follow by recording the change in absorbance at 340 nm resulting from oxidation by malate dehydrogenase.

Photosynthetic Carbon-reduction Cycle Activity

Ribose 5-phosphate/ATP dependent enzymic CO_2 fixation. Incorporation of $^{14}CO_2$ into acid stable products can be determined quite cheaply by incubating enzyme preparations or broken chloroplasts with 100 μmol Tris HCl-pH 8.3; 1 μmol ribose 5-phosphate, 1 μmol ATP; 10 μmol $MgCl_2$; 1 μmol $NaH^{14}CO_3$ in a total volume of 0.5 ml. This assay determines the combined activity of ribose-5-isomerase, phosphate carboxylase. If ribulose-bis-P is available it may be substituted for ribose-5-P and ATP in which case only the carboxylase activity is measured. However, as detailed below accurate determination of this enzyme requires care.

9.12. RIBULOSE-1, 5-BISPHOSPHATE (RuBP) CARBOXYLASE/OXYGENASE

by R. LEEGOOD

RuBP carboxylase is the world's most abundant protein and comprises some 50% of total leaf protein. It catalyses two reactions:

$$CO_2 + RuBP \rightarrow 2\ 3\text{-phosphoglycerate}$$
$$O_2 + RuBP \rightarrow 3\text{-phosphoglycerate}$$
$$+ 2\text{-phosphoglycollate}$$

the former the primary mode of CO_2-assimilation, the latter the first reaction in the photorespiratory pathway. Although its measurement is important, it has long proved one of the most difficult enzymes to assay. This is because it is converted from an inactive to an active form by reaction with CO_2 and Mg^{2+}; inactivation readily occurs in their absence:

$$E(\text{inactive}) + CO_2 \xrightleftharpoons{\text{slow}} E \cdot CO_2 + Mg^{2+}$$
$$\xrightleftharpoons{\text{fast}} E \cdot CO_2 \cdot Mg^{2+}(\text{active})$$

(these reactions occur on a lysine residue at an activator site separate from the catalytic site). The enzyme must therefore be preincubated with CO_2 and Mg^{2+} to measure maximum activity. At the high concentration of CO_2 used in the carboxylase assay (see below), oxygenase activity is low. However, the oxygenase activity may also be measured by O_2 uptake in an O_2 electrode (Lorimer et al., 1977), providing CO_2 is not included in the assay (under these conditions the activated enzyme dissociates rapidly, resulting in a non-linear uptake of O_2).

Close study of Lorimer et al. (1977) (and references therein) is advised before embarking on measurements of RuBP carboxylase activity and particular care should be taken in the design of experiments to measure the enzyme in leaf extracts. An added complication is that the extracted enzyme appears to be cold-inactivated in certain plants, such as wheat.

RuBP Carboxylase Assay

The leaf extract, or the purified enzyme, is applied to a column of Sephadex G–25 equilibrated with 100 mM Tris-HCl, pH 8.6, 20 mM $MgCl_2$, 10 mM $NaHCO_3$ and 1 mM dithiothreitol at room temperature and is eluted in a minimal volume (this removes phenolics, $(NH_4)_2\,SO_4$, etc., and activates the enzyme (10–20 minutes may be necessary)). Alternatively, if a suitable extraction buffer is used (e.g. for chloroplasts), $MgCl_2$ and $NaHCO_3$ can be added afterwards to activate the enzyme (Lilley and Walker, 1974). An aliquot of the activated enzyme is added to the reaction mixture:

100 mM Tris-HCl-NaOH, pH 8.2 (buffer made CO_2-free)

  5 mM dithiothreitol

  20 mM $MgCl_2$

20 mM $NaH^{14}CO_3$ (known specific activity)

0.4 mM RuBP

and the reaction is allowed to proceed for 60 seconds before being stopped by the addition of 2 M HCl (100 μl added to 0.5 ml) and the acid-stable radioactivity determined.

The enzyme may also be determined spectrophotometrically using a cheaper assay (utilizing ribose-5-phosphate) (Lilley and Walker, 1974).

Making CO_2-free Solutions

Distilled water is made CO_2-free by boiling it vigorously for about 20 minutes. The container should then be capped with an outlet containing soda-lime. The buffer, etc., is dissolved in the water and adjusted to about pH 4 with HCl. The solution is purged with N_2 for about 10 minutes. The pH is then adjusted with NaOH or KOH which has been freshly prepared in CO_2-free water. The solution is again purged with N_2 and sealed.

Bibliography and Further Reading

BALDRY, C. W., BUCKE, C. and COOMBS, J. (1970) Effects of some phenol oxidase inhibitors on chloroplasts and carboxylating enzymes of sugar cane and spinach. *Planta* **94**, 124–129.

BARMAN, T. E. (1969) *Enzyme Handbook.* (Springer-Verlag: Heidelberg, London, N.Y.)

BOYER, P. D. (ed.) (1970) *The Enzymes—Student Edition.* (Academic Press: New York.)

DIXON, M. and WEBB, E. C. (1964) *Enzymes*, second edition. (Academic Press: N.Y., London.)

Gelfiltration, Theory and Practice. Pharmacia Fine Chemicals, Box 175, S75104, Uppsala, Sweden.

LILLEY, R. McC. and WALKER, D. A. (1974) An improved spectrophotometric assay for ribulose-bisphosphate carboxylase. *Biochim. Biophys. Acta* **358**, 226–229.

LORIMER, G. H., BADGER, M. R. and ANDREWS, T. J. (1977) D-ribulose-1, 5-bisphosphate carboxylase-oxygenase. Improved methods for the activation and assay of catalytic activities. *Analyt. Biochem.,* **78**, 66–75.

MAHLER, H. R. and CORDES E. H. (1966) *Enzyme Kinetics in Biological Chemistry*, chapter 6. (Harper & Row: N.Y., London.)

WOOD, W. I. (1976) Tables for the preparation of ammonium sulphate solutions. *Analyt. Biochem.* **73**, 250–257.

Appendix A.9.1. Enzyme Kinetics

E = enzyme concentration; S = substrate concentration; T = time; K = constant; K_1, K_2, K_3 = constants; K_m = Michaelis constant; ES = enzyme substrate complex: V = maximum velocity; v = velocity; n = Hill coefficient.

When E is kept constant and S varied between wide limits variation in initial velocity—$\partial S / \partial t$ may be described by a bi-phasic curve:

At low substrate concentrations, the first-order equation, $\partial S / \partial t = K \cdot E \cdot S$., applies and the initial velocity is directly proportional to the initial substrate concentrations. At high substrate concentrations the zero-order equation applies, $(\partial S / \partial T) = K \cdot E$.

Explanation: Theory assumes that E combines with S to form a complex ES then

$$E + S \underset{K_2}{\overset{K_1}{\rightleftharpoons}} ES \xrightarrow{K_3} E + \text{products}.$$

If $E + S \rightleftharpoons ES$ is a reversible process then K_m

(the dissociation constant of ES) $= \dfrac{(E - ES) \cdot S}{ES}$

$$\therefore ES = \frac{E \cdot S}{K_m + S}.$$

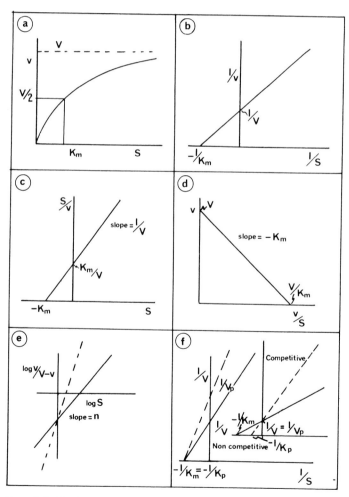

FIG. 9.2. Various kinetic plots: (a) Michaelis–Menton; (b) Lineweaver-Burk; (c) Eadie plot; (d) Hofstee plot; (e) Hill plot; (f) competitive and non-competitive Lineweaver–Burk plots.

If velocity constant for dissociation of $ES = K_3$ then $V \rightarrow K_3 ES = K_3 \cdot E \cdot S / K_m + S$. $v = V$ when ES is maximal, i.e. $ES = E$. Under these circumstances $V = K_3 ES = K_3 E$,

$$\therefore v = \frac{V \cdot S}{K_m + S} \text{ or } K_m = S(V - 1)$$

(Michaelis–Menton)

Hence if $-\log S$ is plotted against v/V a sigmoid curve is obtained with an inflection point at $v/V = \frac{1}{2}$. This point corresponds to a value of $-\log S$ from which K_m may be calculated.

If the reciprocal of the Michaelis–Menton equation is taken, we have the Lineweaver–Burk equation

$$\frac{1}{v} = \frac{K_m + S}{V \cdot S} = \frac{K_m}{V}\left(\frac{1}{S}\right) + \frac{1}{V}.$$

Plot $1/v$ against $1/S$. Intercept at ordinate $= 1/V$ slope of line $= K_m/V$. Alternatively multiply both sides by S then

$$\frac{S}{v} = \frac{K_m}{V} + \frac{S}{V}.$$

Plot S/v against S. Slope is $1/V$ intercept K_m/V or since $v = V - K_m \cdot v/S$, plot v against v/s. The resulting straight line (slope $= -K_m$) has an ordinate intercept of V and an intercept on the abscissa of V/K_m (Hofstee).

The Michaelis–Menton plot may be fitted to a sigmoid curve by taking the empirical form

$$v = \frac{VS^n}{K_m + S^n}$$

hence $\log v/V - v = n \log S - \log K_m = n \log S - C$ (C is a constant). This gives a linear plot of $\log v/V - v$ against $\log S$. This expression is the same as that for the Hill plot.

Examples of the above plots are given in Fig. 9.2.

Inhibition

Only two types of inhibition of enzyme activity will be considered here ($I =$ concentration of inhibitor).

1. Inhibitor competes with substrate for enzyme. If S is high $v_i = V$ ($v_i =$ velocity with inhibitor).
2. Non-competitive
   Inhibition depends only on I.

Hence one must consider not only $E + S = ES$ but also $E + I = EI$.

For competitive inhibitor: Free enzyme $= E - ES - EI$.

$$\therefore K_i = (E - ES - EI)(I)/EI.$$

$$\therefore \text{At steady state } ES = \frac{E \cdot S \cdot K_i}{K_m K_i + K_m I + K_i S}.$$

$$\therefore v = \frac{V \cdot S \cdot K_i}{K_m K_i + K_m I + K_i S}.$$

Modify as Lineweaver–Burk

$$\frac{1}{v} = \frac{1}{V}\left(K_m + \frac{K_m}{K_i} I\right)(1/S) + \frac{1}{V}.$$

Now if $1/v$ plotted against $1/S$ as before, the slope of the resulting straight line is $(K_m/V)(1 + I/K_i)$ and the intercept is $1/V$.

Since K_m may be calculated in the absence of an inhibitor the value of K_i may be calculated.

The effect of a competitive inhibitor is to increase the slope of the line by the factor $1 + I/K_i$ without change in the intercept at $1/V$. Hence if the substrate concentration is large enough the effect of the inhibitor can be overcome.

Appendix A.9.2. The Enzyme Commission Classification System

This may be summarized as follows:

(i) Each enzyme possesses a number made up of four figures.

(ii) The first figure shows to which of the following main groups the enzyme belongs:
Group 1 Oxidoreductases
Group 2 Transferases
Group 3 Hydrolases
Group 4 Lyases (these enzymes catalyse the addition of groups to double bonds or, conversely, remove groups from their substrates leaving double bonds)
Group 5 Isomerases
Group 6 Ligases (synthetases: catalysing the condensation of two molecules coupled with the cleavage of a pyrophosphate bond of ATP or a similar triphosphate)

(iii) The second figure indicates the sub-class and the third figure the sub-sub-class to which the enzyme belongs.

(iv) The fourth figure is the serial number of the enzyme in its subclass.

1. *Oxidoreductases*
1.1 Acting on the CH-OH group of donors
    1.1.1 With NAD or NADP as acceptor
    1.1.2 With a cytochrome as an acceptor
    1.1.3 With O_2 as acceptor
    1.1.99 With other acceptors

1.2 Acting on the aldehyde or keto-group of donors
    1.2.1 With NAD or NADP as acceptor
    1.2.2 With a cytochrome as an acceptor
    1.2.3 With O_2 as acceptor
    1.2.4 With lipoate as acceptor
    1.2.99 With other acceptors

1.3 Acting on the CH-CH group of donors
    1.3.1 With NAD or NADP as acceptor
    1.3.2 With a cytochrome as an acceptor
    1.3.3 With O_2 as acceptor
    1.3.99 With other acceptors

1.4 Acting on the $CH-NH_2$ group of donors
    1.4.1 With NAD or NADP as acceptor
    1.4.3 With O_2 as acceptor

1.5 Acting on the C-NH group of donors
    1.5.1 With NAD or NADP as acceptor
    1.5.3 With O_2 as acceptor

1.6 Acting on reduced NAD or NADP as donor
    1.6.1 With NAD or NADP as acceptor
    1.6.2 With a cytochrome as an acceptor
    1.6.4 With a disulphide compound as acceptor
    1.6.5 With a quinone or related compound as acceptor
    1.6.6 With a nitrogenous group as acceptor
    1.6.99 With other acceptors

1.7 Acting on other nitrogenous compounds as donors
    1.7.3 With O_2 as acceptor
    1.7.99 With other acceptors

1.8 Acting on sulphur groups of donors
    1.8.1 With NAD or NADP as acceptor
    1.8.3 With O_2 as acceptor
    1.8.4 With a disulphide compound as acceptor
    1.8.5 With a quinone or related compound as acceptor
    1.8.6 With a nitrogenous group as acceptor

1.9 Acting on haem groups of donors
    1.9.3 With O_2 as acceptor
    1.9.6 With a nitrogenous group as acceptor

1.10 Acting on diphenols and related substances as donors

1.10.3 With O_2 as acceptor

1.11 Acting on H_2O_2 as acceptor

1.12 Acting on hydrogen as donor

1.13 Acting on single donors with incorporation of oxygen (oxygenases).

1.14 Acting on paired donors with incorporation of oxygen into one donor (hydroxylases)
  1.14.1 Using reduced NAD or NADP as one donor
  1.14.2 Using ascorbate as one donor
  1.14.3 Using reduced pteridine as one donor

2. *Transferases*
  2.1 Transferring one-carbon groups
    2.1.1 Methyltransferases
    2.1.2 Hydroxymethyl-, formyl- and related transferases
    2.1.3 Carboxyl- and carbamoyltransferases
    2.1.4 Amidinotransferases

  2.2 Transferring aldehydic or ketonic residues

  2.3 Acyltransferases
    2.3.1 Acyltransferases
    2.3.2 Aminoacyltransferases

  2.4 Glycosyltransferases
    2.4.1 Hexosyltransferases
    2.4.2 Pentosyltransferases

  2.5 Transferring alkyl or related groups

  2.6 Transferring nitrogenous groups
    2.6.1 Aminotransferases
    2.6.3 Oximinotransferases

  2.7 Transferring phosphorus-containing groups
    2.7.1 Phosphotransferases with an alcohol group as acceptor
    2.7.2 Phosphotransferases with a carboxyl group as acceptor

  2.7.3 Phosphototransferases with a nitrogenous group as acceptor
  2.7.4 Phosphotransferases with a phospho-group as acceptor
  2.7.5 Phosphotransferases, apparently intramolecular
  2.7.6 Pyrophosphotransferases
  2.7.7 Nucleotidyltransferases
  2.7.8 Transferases for other substituted phospho-groups

  2.8 Transferring sulphur-containing groups
    2.8.1 Sulphurtransferases
    2.8.2 Sulphotransferases
    2.8.3 CoA-transferases

3. *Hydrolases*
  3.1 Acting on ester bonds
    3.1.1 Carboxylic ester hydrolases
    3.1.2 Thiolester hydrolases
    3.1.3 Phosphoric monoester hydrolases
    3.1.4 Phosphoric diester hydrolases
    3.1.5 Triphosphoric monoester hydrolases
    3.1.6 Sulphuric ester hydrolases

  3.2 Acting on glycosyl compounds
    3.2.1 Glycoside hydrolases
    3.2.2 Hydrolysing N-glycosyl compounds
    3.2.3 Hydrolysing S-glycosyl compounds

  3.3 Acting on ether bonds
    3.3.1 Thioether hydrolases

  3.4 Acting on peptide bonds (peptide hydrolases)
    3.4.1 α-Amino-acyl-peptide hydrolases
    3.4.2 Peptidyl-amino-acid hydrolases
    3.4.3 Dipeptide hydrolases
    3.4.4 Peptidyl-peptide hydrolases

  3.5 Acting on C—N bonds other than peptide bonds
    3.5.1 In linear amides
    3.5.2 In cyclic amides
    3.5.3 In linear amidines
    3.5.4 In cyclic amidines
    3.5.5 In cyanides
    3.5.99 In other compounds

3.6 Acting on acid-anhydride bonds
   3.6.1 In phosphoryl-containing anhy-drides

3.7 Acting on C—C bonds
   3.7.1 In ketonic substances

3.8 Acting on halide bonds
   3.8.1 In C-halide compounds
   3.8.2 In P—halide compounds

3.9 Acting on P—N bonds

4. *Lyases*
  4.1 Carbon-carbon lyases
    4.1.1 Carboxy-lyases
    4.1.2 Aldehyde-lyases
    4.1.3 Ketoacid-lyases

  4.2 Carbon-oxygen lyases
    4.2.1 Hydro-lyases
    4.2.99 Other carbon-oxygen lyases

  4.3 Carbon-nitrogen lyases
    4.3.1 Ammonia-lyases
    4.3.2 Amidine-lyases

  4.4 Carbon-sulphur lyases

  4.5 Carbon-halide lyases

  4.99 Other lyases

5. *Isomerases*
  5.1 Racemases and epimerases
    5.1.1 Acting on amino acids and derivatives
    5.1.2 Acting on hydroxyacids and derivatives

    5.1.3 Acting on carbohydrates and derivatives
    5.1.99 Acting on other compounds

  5.2 Cis-trans isomerases

  5.3 Intramolecular oxidoreductases
    5.3.1 Interconverting aldoses and ketoses
    5.3.2 Interconverting keto- and enol-groups
    5.3.3 Transposing C=C bonds

  5.4 Intramolecular transferases
    5.4.1 Transferring acyl groups
    5.4.2 Transferring phosphoryl groups
    5.4.99 Transferring other groups

  5.5 Intramolecular lyases

  5.99 Other isomerases

6. *Ligases*
  6.1 Forming C—O bonds
    6.1.1 Amino-acid-RNA ligases

  6.2 Forming C—S bonds
    6.2.1 Acid-thiol ligases

  6.3 Forming C—N bonds
    6.3.1 Acid-ammonia ligases (amide syn-thetases)
    6.3.2 Acid-amino-acid ligases (peptide synthetases)
    6.3.3 Cyclo-ligases
    6.3.4 Other C—N ligases
    6.3.5 C—N ligases with glutamine as N-donor

  6.4 Forming C—C bonds

SECTION 10

BIOMASS FACTS AND FIGURES

by D. O. HALL and J. COOMBS

10.1. FOSSIL FUEL RESERVES AND RESOURCES, BIOMASS PRODUCTION AND CO_2 BALANCES

(Hall, 1979)

Proven reserves	Tonnes coal equivalent
Coal	5×10^{11}
Oil	2×10^{11}
Gas	1×10^{11}
	8×10^{11} t $= 25 \times 10^{21}$ J

Estimated resources	Tonnes coal equivalent
Coal	85×10^{11}
Oil	5×10^{11}
Gas	3×10^{11}
Unconventional gas and oil	20×10^{11}
	113×10^{11} t $= 300 \times 10^{21}$ J
Fossil fuels used so far (1976)	2×10^{11} t carbon $= 6 \times 10^{21}$ J
World's annual energy use	3×10^{20} J
	(5×10^9 t carbon from fossil fuels)

Annual photosynthesis	
(a) net primary production	8×10^{10} t carbon (2×10^{11} t organic matter) $= 3 \times 10^{21}$ J
(b) cultivated land only	0.4×10^{10} t carbon
Stored in biomass	
(a) total (90% in trees)	8×10^{11} t carbon $= 20 \times 10^{21}$ J
(b) cultivated land only (standing mass)	0.06×10^{11} t carbon
Atmospheric CO_2	7×10^{11} t carbon
CO_2 in ocean surface layers	6×10^{11} t carbon
Soil organic matter	$10–30 \times 10^{11}$ t carbon
Ocean organic matter	17×10^{11} t carbon

These data, although imprecise, show that (a) the world's annual use of energy is only one-tenth the annual photosynthetic energy storage, (b) stored biomass on the earth's surface at present is equivalent to the proven fossil fuel reserves, (c) the total stored as fossil carbon only represents about 100 years of net photosynthesis, and (d) the amount of carbon stored in biomass is approximately the same as the atmospheric carbon (CO_2) and the carbon as CO_2 in the ocean surface layers.

10.2. PRIMARY PHOTOSYNTHETIC PRODUCTIVITY OF THE EARTH (*Lieth and Whittaker*, 1975)

	Area (total = 510 million km²)		Net productivity (total = 155.2 billion tons dry wt./yr)	
Total Earth	100		100	
Continents	29.2		64.6	
Forests		9.8		41.6
Tropical Rain		3.3		21.9
Raingreen		1.5		7.3
Summer Green		1.4		4.5
Chaparral		0.3		0.7
Warm Temperate Mixed		1.0		3.2
Boreal (Northern)		2.4		3.9
Woodland		1.4		2.7
Dwarf and Scrub		5.1		1.5
Tundra		1.6		0.7
Desert Scrub		3.5		0.8
Grassland		4.7		9.7
Tropical		2.9		6.8
Temperate		1.8		2.9
Desert (Extreme)		4.7		0
Dry		1.7		0
Ice		3.0		0
Cultivated Land		2.7		5.9
Freshwater		0.8		3.2
Swamp and Marsh		0.4		2.6
Lake and Stream		0.4		0.6
Oceans	70.8		35.4	
Reefs and Estuaries		0.4		2.6
Continental Shelf		5.1		6.0
Open Ocean		65.1		26.7
Upwelling Zones		0.08		0.1

10.3. SOME HIGH SHORT-TERM DRY WEIGHT YIELDS OF CROPS AND THEIR SHORT-TERM PHOTOSYNTHETIC EFFICIENCIES (*Solar Energy—a UK Assessment*)

Crop	Country	g/m²/day	Photosynthetic efficiency (% of total radiation)
Temperate			
Tall fescue	U.K.	43	3.5
Rye-grass	U.K.	28	2.5
Cocksfoot	U.K.	40	3.3
Sugar beet	U.K.	31	4.3
Kale	U.K.	21	2.2
Barley	U.K.	23	1.8
Maize	U.K.	24	3.4
Wheat	Netherlands	18	1.7
Peas	Netherlands	20	1.9
Red clover	New Zealand	23	1.9
Maize	New Zealand	29	2.7
Maize	U.S.A., Kentucky	40	3.4

Crop	Country	g/m²/day	Photosynthetic efficiency (% of total radiation)
Sub-tropical			
Alfalfa	U.S.A., California	23	1.4
Potato	U.S.A., California	37	2.3
Pine	Australia	41	2.7
Cotton	U.S.A., Georgia	27	2.1
Rice	S. Australia	23	1.4
Sugar cane	U.S.A., Texas	31	2.8
Sudan grass	U.S.A., California	51	3.0
Maize	U.S.A., California	52	2.9
Algae	U.S.A., California	24	1.5
Tropical			
Cassava	Malaysia	18	2.0
Rice	Tanzania	17	1.7
Rice	Philippines	27	2.9
Palm oil	Malaysia (whole year)	11	1.4
Napier grass	El Salvador	39	4.2
Bullrush millet	Australia, N.T.	54	4.3
Sugar cane	Hawaii	37	3.8
Maize	Thailand	31	2.7

Other yields: Loomis and Gerakis (see Cooper, 1975) discuss figures for (a) sunflower, growth rates of 79 to 104 g/m²/day have been reported, with a 3-week mean rate of 63.8 g/m²/day giving a photosynthetic efficiency of 7.5%; (b) carrot, growth rates of 146 g/m²/day and a dry matter yield of 54.5 tonnes/ha after 160 days were reported.

Note: Yields in g/m²/day can be converted to tonnes/ha/yr by multiplying by 3.65.

10.4. AVERAGE-TO-GOOD ANNUAL YIELDS OF DRY-MATTER PRODUCTION (*Solar Energy—a UK Assessment*)

	Tonnes/hectare/yr	g/m²/day	Photosynthetic efficiency (percent of total radiation)
Tropical			
Napier grass	88	24	1.6
Sugar cane	66	18	1.2
Reed swamp	59	16	1.1
Annual crops	30	—	—
Perennial crops	75–80	—	—
Rain forest	35–50	—	—
Temperate (Europe)			
Perennial crops	29	8	1.9
Annual crops	22	6	0.8
Grassland	22	6	0.8
Evergreen forest	22	6	0.8
Deciduous forest	15	4	0.6
Savanna	11	3	—
Desert	1	0.3	0.02

HARVESTABLE DRY MATTER ($t\,ha^{-1}\,yr^{-1}$)

(a) Average to Good Annual Yields of Dry-matter Production (Hudson, 1975)

	Typical	High
Sugar cane	35	90
Maize	10	40
Wheat	5	20
Rice	4	16
Sugar beet	8	18
Temperate grass	7	25
Tropical grass	15	50
Cassava	8	35

(b) Maximum Photosynthetic Productivity (U.S.A.)* (Bassham, 1977)

Napier grass	139
Sugar cane	138
Sorghum	186
Sugar beet	113
Alfalfa	84

*All-year growth assumed (often not the case).

Theoretical Maximum

U.S.A. average	224
U.S.A. southwest	263

10.5. AVAILABLE LAND AND CURRENT USE ON A GLOBAL BASIS
(FAO 1978, 1979) (Coombs, 1980)

	Land area 10^6 ha)	Use (%)				
		Permanent crops	Arable	Pasture	Forests	Irrigation
World	13078	0.7	10	23	32	1.5
Developed countries	5484	0.4	12	23	34	0.9
Developing countries	7593	0.9	10	23	30	1.9
U.S.A.	936	0.2	20	26	31	18.4
Canada	922	—	5	3	36	0.05
Europe	472	3.2	30	19	32	2.8
U.S.S.R.	2227	0.2	10	17	41	0.7
Asia	2757	1.0	16	20	22	4.6
S. America	1753	1.2	65	25	53	0.4
Africa	2964	0.5	7	27	22	0.3
Oceania	842	0.1	5	56	22	0.2

10.6. AREAS OF CLOSED FOREST AND SIZE OF THE GROWING STOCK (Coombs, 1980)

	Closed forest (10^6 ha)	Growing stock of closed forests		
		Total (10^6m^3)	m^3/ha	m^3/caput
U.S.A.	220	20,200	92.1	95.3
Canada	250	17,800	71.2	809.1
Europe	144	14,900	103.5	29.1
U.S.S.R.	770	81,800	106.2	324.6
Asia and Oceania	448	38,700	72.9	18.7
S. America	631	92,000	145.8	362.2
Africa	188	35,200	187.2	90.0

10.7. TOTAL PRODUCTION (10⁶t) AND YIELDS (t/ha) OF THE MAJOR SUGAR AND STARCH CROPS
(Coombs, 1980)

	(a) Sugar				(b) Starch grains								(c) Starch roots and tubers							
	Cane		Beet		Cereals		Maize		Rice		Wheat		Roots and Tubers		Potatoes		Sweet potatoes		Cassava	
	Total	Yield	Total	Yield	Total	Yield	Total	Yield	Total	Yield	Total	Yield	Total	Yield	Total	Yield	Total	Yield	Total	Yield
World	737	56	290	32	1459	1.9	349	3.0	366	2.6	386	1.7	570	11.0	272	15.0	138	9.6	110	8.8
Developed countries	70	79	263	32	765	2.5	236	4.7	26	5.4	261	1.9	225	15.7	222	15.7	2	14.7	—	—
Developing Countries	667	54	26	30	687	1.6	113	1.7	340	2.5	125	1.3	345	9.2	70	10.4	136	9.5	110	8.8
U.S.A.	25	82	23	46	273	4.1	161	5.7	5	4.9	55	2.0	17	28.0	16	29.2	1	12.4	—	—
Canada	—	—	1	39	42	2.3	48	5.9	—	—	19	1.9	3	22.4	3	22.4	—	—	—	—
Europe	0.3	63	143	38	249	3.5	49	4.2	1.5	3.8	82	3.3	114	18.7	83	11.8	—	—	—	—
U.S.S.R.	—	—	93	25	187	1.5	11	3.2	2.2	4.0	92	1.5	83	10.3	83	11.8	—	—	—	—
Asia	305	52	25	31	603	1.8	54	2.0	335	1.9	108	1.3	224	11.0	61	11.2	128	9.9	33	11.2
S. America	187	57	—	—	64	1.7	31	1.8	13	1.9	9	1.1	44	11.0	9	9.5	3	10.4	32	11.6
Africa	60	64	16	32	66	0.9	26	1.3	8	1.8	8	0.9	73	6.8	4	7.8	5	6.3	44	6.6
Oceania	26	75	—	—	16	1.0	0.4	4.5	0.5	5.3	10	0.9	2	10.3	1	23.0	0.6	5.4	2	11.0

These figures are on a crop basis excluding the considerable amounts of crop wastes, straw, etc. The cereal figures exclude such crops grown for forage silage. (FAO, 1978, 1979).

10.8. PHOTOSYNTHETIC EFFICIENCY AND ENERGY LOSSES (*Solar Energy, a UK Assessment*)

	Available light energy
At sea level	100%
50% loss as a result of 400–700 nm light being the photosynthetically usable wavelengths	50%
20% loss due to reflection, absorption and transmission by leaves	40%
77% loss representing quantum efficiency requirements for CO_2 fixation in 680 nm light (assuming 10 quanta/CO_2)[a] and that the energy content of 575 nm red light is the radiation peak of visible light	9.2%
40% loss due to respiration	5.5%
	Overall PS efficiency

[a] If the minimum quantum requirement is 8 quanta/CO_2, then this loss factor becomes 72% (instead of 77%) giving a final photosynthetic efficiency of 6.7% (instead of 5.5%).

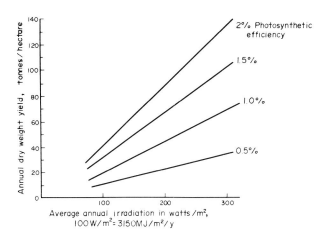

Expected annual plant yields as a function of annular solar irradiation at various photosynthetic efficiencies

10.9. SOURCE OF BIOMASS FOR CONVERSION TO FUELS
(Coombs, 1980)

Wastes	Land crops	Aquatic plants
Manures	*Ligno-cellulose*	*Algae*
Slurry	Trees	Unicellular
Domestic rubbish	Eucalyptus	Chlorella
Food wastes	Poplar	Scenedesmus
Sewage	Firs, Pines	Navicula
	Luceana, etc.	Multicellular
		Kelp
Residues		
Wood residues	*Starch crops*	*Water weed*
Cane tops	Maize	Water hyacinth
Straw	Cassava	*Water reeds/rushes*
Husks		
Citrus peel	*Sugar crops*	
Bagasse	Cane	
Molasses	Beet	

10.10. CONVERSION TECHNIQUES FOR PRODUCING FUELS FROM BIOMASS
(Hall and Coombs, 1979)

Process	Products	State of art
(a) Thermal		
Combustion	CO_2, water, ash, heat, steam, electricity	Well established
Pyrolysis	Char, oil, gas (CO, C_2H_4, H_2)	Small scale established
Gasification	Mainly CO and H_2	Large scale under development
Gasification + catalytic synthesis	Methanol, hydrocarbons	
(b) Biological		
Yeast fermentation	Ethanol + CO_2	Established
Anaerobic digestion	Methane + CO_2	Established

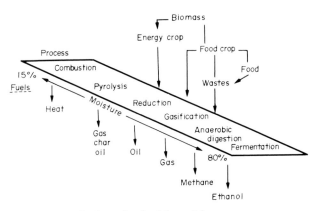

Options for fuel from biomass.

Bibliography and Further Reading

BASSHAM, J. A. (1977) Increasing crop production through more controlled photosynthesis. *Sci* **197**, 630.

Biomass for Energy (1979) Proc. Conference C20, UK-ISES, 19 Albemarle Street, London, U.K. 99 pp.

COOMBS, J. (1980) Renewable sources of energy (carbohydrates). *Outlook on Agriculture* **10**, 235.

COOPER, J. P. (ed.) (1975) *Photosynthesis and Productivity in Different Environments.* (Cambridge University Press, U.K. 715 pp.)

Energy from the Biomass (1979) Watt Committee on Energy, 1 Birdcage Walk, London, SW1, U.K. 76 pp.

HALL, D. O. and COOMBS, J. (1979) *Ibid.,* p. 2.

HALL, D. O. (1980) Renewable resources (hydrocarbons). *Outlook on Agriculture* **10**, 246.

HUDSON, J. C. (1975) Sugar cane: Energy relationships with fossil fuel. *The Sugar Journal* **37**, 25.

LIETH, H. and WHITTAKER, R. H. (eds.) *Primary Productivity of the Biosphere* (1975). (Springer: Berlin. 339 pp.)

PALZ, W., CHARTIER, P., HALL, D. O. (eds.) (1981) *Energy from Biomass.* (Applied Science, Publishers, London, U.K. pp. 984.)

SLESSER, M. and LEWIS, C. (1979) *Biological Energy Resources.* (Spon: London, U.K. 192 pp.)

Solar Energy—a UK Assessment (1976) Published by U.K. Section of the International Solar Energy Society, London. 373 pp.

CONVERSION FACTORS

Units and Conversion Factors

Gigajoule (GJ) = 10^9 joules

Tonne wood (air dry) = 15 GJ

Tonne wood (bone dry) = 20 GJ

Tonne agricultural residues = 13 GJ

Tonne dung = 15 GJ

Tonne charcoal = 30 GJ

Tonne wood = 1.4 m^3 wood

1 m^3 wood = 0.73 tonne

1 m^3 wood = 10 GJ

1 m^3 wood = 0.33 tonne coal equivalent (tce)

12 tonne wood → produce 1 tonne charcoal (by earth kiln)

1 barrel oil = 6.3 GJ

1 tonne oil = 44.7 GJ = 7.1 barrels

1 tonne coal = 28 GJ

1 tonne coal = 1.7 tonne wood = 2.3 m^3 wood

1 GJ = 280 kWh

1 kWh = 3.600×10^6 J

1 GJ = 0.95×10^6 Btu

1 GJ = 0.24×10^6 kcalories

1 calorie = 4.187 J

1 GJ = 26 litres kerosene

1 litre gasoline = 3.5×10^7 J

SOLAR RADIATION ON THE EARTH

The energy from the sun, per unit time, received on a unit area of surface perpendicular to the radiation, in the outer space around the earth above the atmosphere, varies with time of year from about 1.31 to 1.4 kW/m^2 according to the variations in earth–sun distance. The solar constant is the value for the average earth–sun centre-to-centre distance (1.495×10^8 km). The standard value of the solar constant is:

$$E_{sc} = 1.353 \text{ kW/m}^2$$
$$= 4871 \text{ kJ/m}^2 \cdot \text{hr}$$
$$= 1.940 \text{ cal/cm}^2 \cdot \text{min}$$
$$= 429 \text{ Btu/ft}^2 \cdot \text{hr}$$
$$= 1.940 \text{ Langleys/min}$$

The solar constant E_{sc} is the energy received in the space above the atmosphere. Some portion of this energy is reflected back into the outer space from the atmosphere and clouds, and some is absorbed and scattered by molecules and particles in the atmosphere.

Quantities of solar energy received by different areas on the earth's surface
when the solar radiation is 1 kW/m^2

	kWt	kcal/min	Btu/hr	kJ/day*	kcal/day*
1 cm^2	10^{-4}	1.43×10^{-3}	0.34	2.88	0.69
1 m^2	1	14.34	3414	28 800	6883
1 km^2	10^6	1.43×10^7	3.41×10^9	2.88×10^{10}	6.88×10^9
1 in^2	6.45×10^{-4}	9.25×10^{-3}	2.20	18.58	4.44
1 ft^2	9.29×10^{-2}	1.33	317	2676	639
1 mile^2	2.59×10^6	3.71×10^7	8.84×10^9	7.46×10^{10}	1.78×10^{10}

*Assuming 8 hours per day of solar radiation.

INDEX